# Selected Titles in This Series



*(Continued in the back of this publication)*

# DIMACS
Series in Discrete Mathematics
and Theoretical Computer Science

Volume 55


# Discrete Mathematical Problems with Medical Applications

DIMACS Workshop
Discrete Mathematical Problems
with Medical Applications
December 8–10, 1999
DIMACS Center


Ding-Zhu Du
Panos M. Pardalos
Jie Wang
Editors



NSF Science and Technology Center
in Discrete Mathematics and Theoretical Computer Science
A consortium of Rutgers University, Princeton University,
AT&T Labs–Research, Bell Labs (Lucent Technologies),
Telcordia Technologies, and NEC Research Institute

American Mathematical Society

This DIMACS volume contains papers from a DIMACS workshop on Discrete Mathematical Problems with Medical Applications, held December 8–10, 1999.

2000 *Mathematics Subject Classification.* Primary 05B20, 06B99, 52C99, 62K99, 62N01, 68Q17, 68R05, 68R10, 68T05, 68T10, 68T45, 68W40, 68W99, 92C50, 92C55; Secondary 05D05, 15A18, 57N99, 62H99, 62P10, 68Q25, 90C11, 90C30, 90C90.

---


**Library of Congress Cataloging-in-Publication Data**

DIMACS Workshop Discrete Mathematical Problems with Medical Applications (1999 : DIMACS Center)

Discrete mathematical problems with medical applications : DIMACS Workshop Discrete Mathematical Problems with Medical Applications, December 8–10, 1999, DIMACS Center / Ding-Zhu Du, Panos M. Pardalos, Jie Wang, editors.

p. cm. — (DIMACS series in discrete mathematics and theoretical computer science ; v. 55)

Includes bibliographical references.

ISBN 0-8218-2096-6 (alk. paper)

1. Medicine—Mathematics—Congresses. 2. Computer science—Mathematics—Congresses. 3. Biomathematics—Congresses. 4. Mathematics—Congresses. I. Du, Dingzhu. II. Pardalos, P. M. (Panos M.), 1954– III. Wang, Jie, 1961– IV. Title. V. Series.

R853.M3 .D56 1999

610′.1′51—dc21 00-045118


---




10 9 8 7 6 5 4 3 2 1 05 04 03 02 01 00

# Contents

# Foreword

The DIMACS Workshop on Discrete Mathematical Problems with Medical Applications was held Dec. 8–10, 1999 at DIMACS, Rutgers University. We would like to express our appreciation to Ding-Zhu Du, Panos M. Pardalos, and Jie Wang for their efforts to organize and plan this successful conference.

The workshop was part of the Special Year on Mathematical Support for Molecular Biology. We extend our thanks to Martin Farach-Colton, Joachim Messing, Fred Roberts, Larry Shepp, and Mike Waterman for their work as special year organizers.

The conference brought together theoreticians and practitioners working on "discrete mathematical medicine" from a variety of perspectives ranging from medical imaging to DNA screening, with emphasis on optimization and discrete mathematical modeling.

DIMACS gratefully acknowledges the generous support that makes these programs possible. Thanks go to special year supporters: The National Science Foundation, the New Jersey Commission on Science and Technology, AT&T Labs, Bell Labs, BIOSYM Tech, the Centers for Disease Control, Ciba-Geigy Corporation, Department of Energy, International Union of Crystallography, Merck Research, National Center for Human Genome Research, National Institute of Allergy and Infectious Diseases, National Institutes of Health, National Security Agency, Princeton University, Program in Mathematics and Molecular Biology at Florida State University, Promega Corporation, Roche Molecular Systems, Rutgers University, Sandia National Laboratories, the Alfred P. Sloan Foundation, SmithKline Beecham, Telcordia Technologies, and the University of Pennsylvania.

Fred S. Roberts
Director

Robert Sedgewick
Co-Director for Princeton

# Preface

In the context of the 1994-2000 DIMACS special years on **Mathematical Support for Molecular Biology**, a three-day workshop entitled "Discrete Mathematical Problems with Medical Applications" was held on Dec. 8–10, 1999 at DIMACS, at Rutgers University. More than 40 researchers from universities, institutes, governmental agencies, and industrial companies attended the workshop. Participants from different countries gave the workshop an important international component.

The study of discrete mathematics and optimization with medical applications has emerged as an interesting new research area. Significant applications of this research have been found on medical imaging, radiosurgical treatment planning, radiation therapy, virtual endoscopy, and human anatomy modeling, to name a few. This volume represents a selection of refereed papers presented at the workshop.

The papers published here are a cross section of various research topics currently being explored. They range from medical imaging to human anatomy modeling, from gamma knife treatment planning to radiation therapy, from epileptic seizures to DNA screening.

The workshop was sponsored by DIMACS, through grants from the National Science Foundation, the New Jersey Commission on Science and Technology, and other sources. We would like to thank the staff members at DIMACS for helping us to organize and host this workshop. We would also like to thank the participants of the workshop, the contributing authors, the anonymous referees, and Shirley Hill of AMS for helping with this event and the preparation of this volume.

Ding-Zhu Du, Panos M. Pardalos, and Jie Wang
May 2000

DIMACS Series in Discrete Mathematics
and Theoretical Computer Science
Volume **55**, 2000

# Breast Cancer Survival and Chemotherapy: A Support Vector Machine Analysis

Y.-J. Lee, O. L. Mangasarian, and W. H. Wolberg

ABSTRACT. A linear support vector machine (SVM) is used to extract 6 features from a total of 31 features in a dataset of 253 breast cancer patients. Five features are nuclear features obtained during a non-invasive diagnostic procedure while one feature, tumor size, is obtained during surgery. The linear SVM selected the 6 features in the process of classifying the patients into node-positive (patients with some metastasized lymph nodes) and node-negative (patients with no metastasized lymph nodes). Node-positive patients are typically those who undergo chemotherapy. The 6 features were then used in a Gaussian kernel nonlinear SVM to classify the patients into three prognostic groups: good (node-negative), intermediate (1 to 4 metastasized nodes) and poor (more than 4 metastasized nodes). Very well separated Kaplan-Meier survival curves were constructed for the three groups with pairwise p-value of less than 0.009 based on the logrank statistic. Patients in the good prognostic group had the highest survival, while patients in the poor prognostic group had the lowest. The majority (72.8%) of the good group did not receive chemotherapy, while the majority (87.5%) of the poor group received chemotherapy. Just over half (56.7%) of the intermediate group received chemotherapy. New patients can be assigned to one of these three prognostic groups with its associated survival curve, based only on 6 features obtained before and during surgery, but without the potentially risky procedure of removing lymph nodes to determine how many of them have metastasized.

## 1. Introduction

Support vector machines [**Vap95, CM98**] as well as linear programming approaches [**Man65, Man68, BM92**] have been extensively used in machine learning and data mining applications [**Man97, BMS98, BM98a, BM98b, BFM99**] as well as in medical applications [**MSW95, WSM93, WSHM95b, WSHM95a, WSM97**]. In this work we attempt to classify breast cancer patients using a criterion that is closely related to the decision whether a patient is prescribed to have chemotherapy treatment or not. This criterion is the presence of metastasized lymph nodes under the patient's armpits (node-positive) or their absence

---

2000 *Mathematics Subject Classification.* Primary 68T10, 62N01; Secondary 90C90.

The research described in this Data Mining Institute Report 99-10, December 1999, was supported by National Science Foundation Grants CCR-9729842 and CDA-9623632, by Air Force Office of Scientific Research Grant F49620-97-1-0326 and by the Microsoft Corporation.

(node-negative). Lymph nodes are removed during surgery in conjunction with the removal of the malignant tumor from the breast. This potentially risky procedure which can cause arm swelling and increased susceptibility to infection can possibly be eliminated by using the classification procedures proposed here. By using a linear support vector machine we first select 6 out of 31 available features to classify patients into node-positive and node-negative patients. We then use these six features in a nonlinear support vector machine to classify 253 breast cancer patients into three prognosis groups: a good prognosis group (GPG) which is node-negative, an intermediate prognosis group (IPG) which has 1-4 metastasized lymph nodes, and a poor prognosis group (PPG) which has more than 4 metastasized lymph nodes. It turns out that these three groups have well separated survival curves [**KM58, Kle96**] with very small inter-curve p-values (Figure 3). In addition, 72.8% of the patients in GPG did not take chemotherapy, while 87.5% of the patients in PPG took chemotherapy. Only 56.7% of patients in IPG took chemotherapy. Thus our classification procedure can be utilized to assign new patients to one of three prognostic groups with an associated survival curve and a possible indication of the utilization of chemotherapy or not.

The paper is organized as follows. Section 2 gives a basic description of the essentials of support vector machines. Section 3 applies the ideas described in Section 2 to the problem of classifying patients using lymph node status as a classification criterion. Section 4 gives some brief conclusions.

A word about our notation now. All vectors will be column vectors unless transposed to a row vector by a prime superscript $'$. The scalar (inner) product of two vectors $x$ and $y$ in the $n$-dimensional real space $R^n$ will be denoted by $x'y$. For a general norm $\|\cdot\|$ on $R^n$, the dual norm $\|\cdot\|'$ on $R^n$ is defined as

$$\|x\|' := \max_{\|y\|=1} x'y. \tag{1.1}$$

For $p, q \in [1, \infty]$, $\frac{1}{p} + \frac{1}{q} = 1$, the $p$-norm $\|\cdot\|_p$ and $q$-norm $\|\cdot\|_q$ are dual norms. Thus the 1-norm $\|\cdot\|_1$ and $\infty$-norm $\|\cdot\|_\infty$ are dual norms while the 2-norm is self-dual. For an $\ell \times n$ matrix $A$, $A_i$ will denote the $i$th row of $A$. The identity matrix in a real space of arbitrary dimension will be denoted by $I$, while a column vector of ones of arbitrary dimension will be denoted by $e$. We shall employ the MATLAB "dot" notation [**MAT92**] to signify application of a function to all components of a matrix or a vector. For example if $A \in R^{\ell \times n}$, then $A^2_{\bullet} \in R^{\ell \times n}$ will denote the matrix obtained by squaring each element of $A$.

## 2. Support Vector Machines

We begin with the simple linear support vector machine formulation as follows:

$$\begin{array}{rll} \min\limits_{w,\gamma,y} & \nu e'y + \|w\| & \\ \text{s.t.} & D(Aw - e\gamma) + y & \geq e \\ & y & \geq 0. \end{array} \tag{2.1}$$

Here, $\nu$ is a positive weight, $\|\cdot\|$ is an arbitrary norm and the $m \times n$ matrix $A$ represents $m$ given points in $R^n$ which belong to class 1 or -1 depending on whether the corresponding elements of the given $m \times m$ diagonal matrix $D$ are 1 or -1 respectively. The plane in $R^n$, with normal $w$ and distance from the origin $\frac{|\gamma|}{\|w\|}$

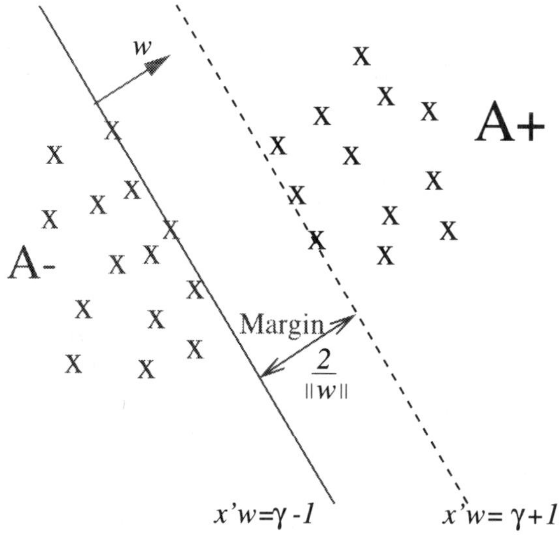


FIGURE 1. A separating plane $x'w = \gamma$, midway and parallel the bounding parallel planes $x'w = \gamma \pm 1$ with maximal separation margin $\frac{2}{\|w\|}$, is obtained by solving the mathematical program (2.1).

from the origin, generated by a solution $(w, \gamma)$ of the above mathematical program:

$$x'w = \gamma, \tag{2.2}$$

strictly separates the set $A+ := \{A_i \mid D_{ii} = 1\}$ from $A- := \{A_i \mid D_{ii} = -1\}$, if and only if the slack variable $y \in R^m$ is zero as shown in Figure 1. The margin or distance (measured by a norm $\|\cdot\|'$ that is dual to the norm $\|\cdot\|$ of (2.1)) between the bounding planes $x'w = \gamma \pm 1$ is given by $\frac{2}{\|w\|}$ [**Man99**]. These two planes are parallel to the plane $x'w = \gamma$ and bound the sets $A+$ and $A-$. For $\nu \geq \bar{\nu}$ for some $\bar{\nu} > 0$ [**MM79**, Theorem 4], the margin is maximized by a solution of (2.1) that suppresses the slack variable $y$ first. When the sets $A+$ and $A-$ are not strictly linearly separable, some components $y_i$ of the slack variable $y$ will be positive. The points $A_i$ in either $A+$ or $A-$ corresponding to positive $y_i$, as well as those corresponding to zero $y_i$ but with positive Lagrange multipliers, are termed *support vectors*. Points $A_i$ with positive $y_i$ lie on the wrong sides of the bounding planes $x'w = \gamma \pm 1$, while points with positive multipliers and zero slacks $y_i$ lie on the bounding planes.

We shall employ the 1-norm in the SVM formulation which leads to the following linear programming formulation which has been shown [**BM98a**] to have one of the best feature selection properties among all norms including the conventional 2-norm-squared formulation [**Vap95, CM98**]:

$$\begin{array}{rcrcl} \min\limits_{w,\gamma,s,y} & & \nu e'y + e's & & \\ \text{s.t.} & & D(Aw - e\gamma) + y & \geq & e \\ & -s \leq & w & \leq & s \\ & & y & \geq & 0. \end{array} \tag{2.3}$$

This linear programming formulation will be our *SVM1* formulation that will be used for feature selection in the next section of the paper.

In order to obtain a nonlinear support vector machine with a nonlinear separating surface, we make use of the following transformation based on the dual formulation for the 2-norm-squared support vector machine [**Man00**]:

$$w = A'Du \tag{2.4}$$

Substitution in (2.1) gives:

$$\begin{array}{lrcl} \min\limits_{u,\gamma,y} & \nu e'y + \|A'Du\| & & \\ \text{s.t.} & D(AA'Du - e\gamma) + y & \geq & e \\ & y & \geq & 0. \end{array} \tag{2.5}$$

Replacing $AA'$ by a nonlinear kernel $K(A, A') : R^{m\times n} \times R^{n\times m} \longrightarrow R^{m\times m}$, and $\|A'Du\|$ by a convex function of $u$, typically a norm, results in the following nonlinear generalized support vector machine (GSVM) [**Man00**]:

$$\begin{array}{lrcl} \min\limits_{u,\gamma,y} & \nu e'y + \|u\|_p & & \\ \text{s.t.} & D(K(A, A')Du - e\gamma) + y & \geq & e \\ & y & \geq & 0, \end{array} \tag{2.6}$$

with the following nonlinear separating surface instead of a plane:

$$K(x', A')Du = \gamma. \tag{2.7}$$

Typical kernel functions are the following:

- **Polynomial Kernel** $(AA' + \mu aa')^d_{\bullet}$,
  where $[\cdot]^d_{\bullet}$ denotes component-wise exponentiation as in MATLAB [**MAT92**].
- **Radial Basis (Gaussian) Kernel** $\varepsilon^{-\mu\|A_i - A_j\|^2}$, $i, j = 1, \ldots, m$.
- **Neural Network Kernel** $(AA' + \mu aa')_{\bullet *}$,
  where $[\cdot]_{\bullet *}$ denotes the step function: $R \longrightarrow \{0, 1\}$ component-wise.

Using the 1-norm in (2.6) leads to a linear programming formulation of the nonlinear GSVM that we will be using in this work:

$$\begin{array}{lrcl} \min\limits_{u,\gamma,s,y} & & \nu e'y + e's & & \\ \text{s.t.} & & D(K(A, A')Du - e\gamma) + y & \geq & e \\ & -s \leq & u & \leq & s \\ & & y & \geq & 0. \end{array} \tag{2.8}$$

This nonlinear GSVM, which is capable of discriminating between elements of fairly complex sets, such as the black and white squares of a checkerboard [**MM99, LM99**], will be used in the next section. We note that even though the nonlinear GSVM (2.8) is capable of achieving fairly complex separation, it is not suited for feature selection, because the kernel $K(A, A')$ creates interactions between all the features through the variable $u$ which is not present in SVM1 (2.3). That is the reason we used the linear SVM1 for feature selection and the nonlinear GSVM for the final separation.

## 3. Breast Cancer Prognosis and Chemotherapy

In this section of the paper we shall use the support vector machines described in the previous section to discriminate between 253 breast cancer patients. The dataset for these patients is publicly available [**WLM99**]. Our criterion for discrimination will be the absence or presence of metastasized lymph nodes (1 to 30 nodes) removed from the patient's armpit during surgery. Lymph node metastasis is a strong indicator for chemotherapy. The total number of features used to constitute the $n$-dimensional space in which the separation is accomplished is made up of the mean, standard error, and maximum (worst) value of ten cytological nuclear measurements of size, shape and texture taken from the patients breast by a non-surgical fine needle aspirate procedure [**Str94, MSW95**] together with the tumor size excised from the patient's breast during surgery. Taken together these constitute the 31-dimensional feature space over which classification according to lymph node metastasis will be achieved.

Before proceeding with our classification process we exhibit Kaplan-Meier survival curves [**Kle96**] for 140 of the 253 patients who had chemotherapy and for the remaining 113 patients that did not have chemotherapy in Figure 2. The solid curve gives the percent of surviving patients as a function of time for the group that received chemotherapy, while the dashed curve gives the corresponding survival curve for the group that did not receive chemotherapy. One possible interpretation of this curve is that the patients not receiving chemotherapy and who have a better survival curve than those who are receiving chemotherapy, are not being compromised by not receiving chemotherapy. Another interpretation is that most patients who had chemotherapy were preselected for having lymph node metastasis and hence had inherently worse cancers. The p-value based on the logrank statistic [**Kle96**] relating these two curves is 0.0018. Hence these two curves are statistically distinct. (The logrank statistic is a comparative measure of the survival times of two groups which calculates a weighted sum of the difference, at each time interval, of the mortality ratio between the two groups: $\frac{m_1}{m_2}$, and the population ratio between the two groups: $\frac{n_1}{n_2}$. Here, $m_i$ is the number of deaths in group $i$ and $n_i$ is the number of people in group $i$. The logrank statistic for two groups has a Chi-squared distribution with degree of freedom one [**Kle96**]. Larger logrank statistic indicates significantly different survival curves. That corresponds to a small p-value. The p-value is the probability that the two survival curves are statistically the same.)

Our next result is the classification of the 253 patients into three prognostic groups as follows:

1. **Good Prognosis** "Node-negative patients": Patients with no metastasized lymph nodes.
2. **Intermediate Prognosis** "Node-positive patients" with 1 to 4 metastasized lymph nodes.
3. **Poor Prognosis** "Node-positive patients" with more than 4 metastasized lymph nodes.

To achieve this classification we first used SVM1 (2.3) to separate node-negative patients from node-positive patients. By using a sufficiently small value of $\nu$ in (2.3), $\nu = 0.02$, the SVM1 formulation (2.3) suppressed 25 of the 31 features and selected a total of 6 features as follows:

1. **Mean Area**
2. **Standard Error of Area**
3. **Worst Area**
4. **Worst Texture**
5. **Worst Perimeter**
6. **Tumor Size** (from surgery)

Note that the first five are cytological features obtained from the fine needle aspirate taken during diagnosis.

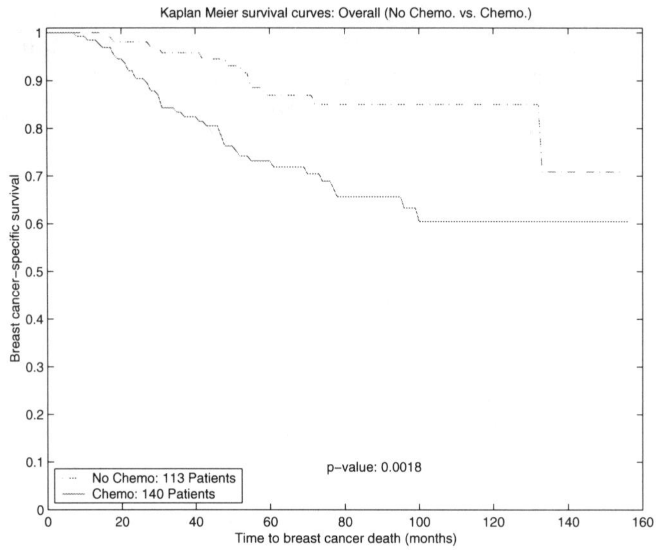


FIGURE 2. Kaplan-Meier survival curve (solid line) for 140 of the 253 patients who received chemotherapy and the corresponding survival curve (dashed line) for the remaining 113 patients that did not receive chemotherapy. p-value=0.0018.

We next used the nonlinear support vector machine GSVM (2.8) with a Gaussian kernel with $\nu = 0.096$ to achieve our three-class separation as follows:

1. **GPG**: Good prognosis group (0 lymph node metastasis)
   was separated from all the rest for which:
2. **IPG**: Intermediate prognosis group (1-4 lymph node metastasis)
   was separated from:
   **PPG**: Poor prognosis group (lymph node metastasis $> 4$).

Note that the separation in Step 2 above included some node-negative patients in IPG because the separation of Step 1 is not perfect and GPG does not include *all* node-negative patients. Tenfold cross validation correctness [**Sto74**] for the above separation was 72.8%.

Based on the above separation we constructed three Kaplan-Meier survival curves for each of the three prognostic groups: GPG, IPG and PPG as shown in Figure 3. Note that the three curves are very well separated and the p-values between the three groups are as follows:

1. GPG-IPG p-value=0.0090
2. GPG-PPG p-value=9.5769e-06
3. IPG-PPG p-value=0.0037

These p-values are substantially smaller than the conventional 0.05 cutoff used to reject the null-hypothesis (survival curves are the same). Hence these curves are indeed statistically significantly different from each other.

It is worth noting within each of the three prognosis groups how many patients received and did not receive chemotherapy:

1. **GPG**: 19 patients received chemotherapy; 51 did not (27.2%;72.8%)
2. **IPG**: 72 patients received chemotherapy; 55 did not (56.7%;43.3%)

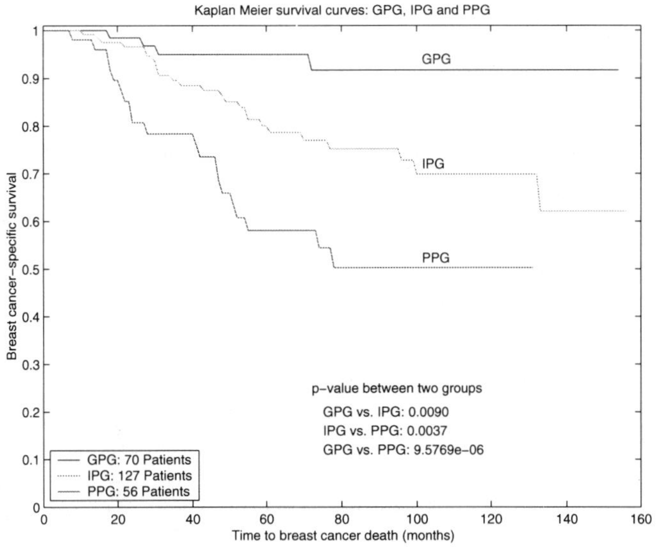


FIGURE 3. Kaplan-Meier survival curves for GPG, IPG and PPG: the good, intermediate and poor prognosis groups. The classification criterion is based on the number of metastasized lymph nodes. The features used for the classification are: mean area, standard error of area, worst texture, worst perimeter and tumor size. The pairwise p-values are: GPG-IPG= 0.0090, GPG-PPG= 9.5769e-06, IPG-PPG= 0.0037.

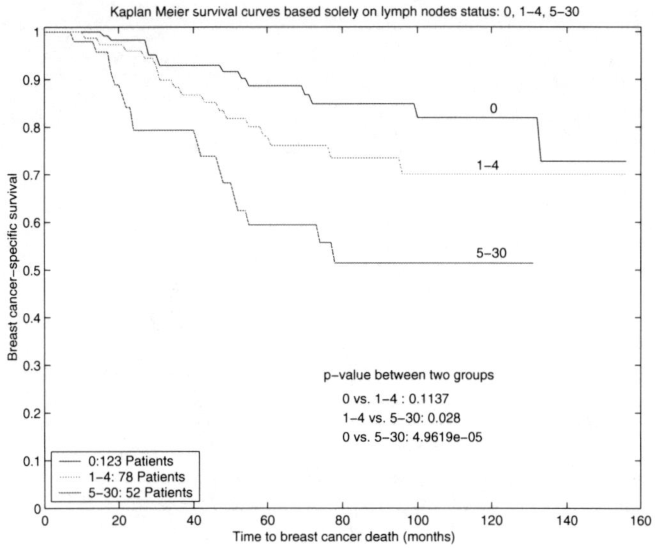


FIGURE 4. Kaplan-Meier survival curves for 3 groups based solely on lymph node status: 0, [1-4] and [5-30]. The pairwise p-values are: [0]-[1-4]:0.1137, [1-4]-[5-30]:0.028, [0]-[5-30]:4.9619e-05.

3. **PPG**: 49 patients received chemotherapy; 7 did not (87.5%;12.5%)

Within each prognosis group, there was no statistically significant difference based on p-values, between the survival curves of patients who took and did not take chemotherapy.

In contrast to the three survival curves based on the GPG, IPG and PPG groups, we constructed three other survival curves based solely on the number of metastasized lymph nodes, the conventional indicator for chemotherapy prescription, as shown in Figure 4. It is interesting to observe the large p-values between the node-negative survival curve and the 1-4 lymph node survival curve (p-value=0.1137), and between the latter curve and the 5-30 lymph node survival curve (p-value=0.028). This indicates that our criteria for GPG-IPG-PPG classification based on the six features selected by SVM1, which do not include lymph node status, yield statistically more distinct survival curves than those based on lymph node status by itself.

## 4. Conclusions

Based on the results presented in this work, the following conclusions can be made:

- Support vector machines can be used to:
  - Select features that are important for prognosis.
  - Classify breast cancer patients into three groups with well separated survival curves.
- Most (72.8%) patients in good group received no chemotherapy.
- Most (87.5%) patients in poor group received chemotherapy.
- Just over half the patients (56.7%) in intermediate group received chemotherapy.
- Survival curves and possible chemotherapy decision for new patients can be determined by assigning them to one of the three groups based on five nuclear features and tumor size without the need for lymph node status.

## References

[BFM99] P. S. Bradley, Usama M. Fayyad, and O. L. Mangasarian, *Data mining: Overview and optimization opportunities*, INFORMS Journal on Computing **11** (1999), 217–238, ftp://ftp.cs.wisc.edu/math-prog/tech-reports/98-01.ps.

[BM92] K. P. Bennett and O. L. Mangasarian, *Robust linear programming discrimination of two linearly inseparable sets*, Optimization Methods and Software **1** (1992), 23–34.

[BM98a] P. S. Bradley and O. L. Mangasarian, *Feature selection via concave minimization and support vector machines*, Machine Learning Proceedings of the Fifteenth International Conference(ICML '98) (San Francisco, California) (J. Shavlik, ed.), Morgan Kaufmann, 1998, ftp://ftp.cs.wisc.edu/math-prog/tech-reports/98-03.ps, pp. 82–90.

[BM98b] P. S. Bradley and O. L. Mangasarian, *Massive data discrimination via linear support vector machines*, Tech. Report 98-05, Computer Sciences Department, University of Wisconsin, Madison, Wisconsin, May 1998, *Optimization Methods and Software*, to appear. ftp://ftp.cs.wisc.edu/math-prog/tech-reports/98-03.ps.

[BMS98] P. S. Bradley, O. L. Mangasarian, and W. N. Street, *Feature selection via mathematical programming*, INFORMS Journal on Computing **10** (1998), no. 2, 209–217, ftp://ftp.cs.wisc.edu/math-prog/tech-reports/95-21.ps.

[CM98] V. Cherkassky and F. Mulier, *Learning from data - concepts, theory and methods*, John Wiley & Sons, New York, 1998.

[Kle96] David G. Kleinbaum, *Survival analysis*, Springer-Verlag, New York, 1996.

[KM58] E. L. Kaplan and P. Meier, *Nonparametric estimation from incomplete observations*, Journal of the American Statistical Association **53** (1958), 457–481.

[LM99] Yuh-Jye Lee and O. L. Mangasarian, *SSVM: A smooth support vector machine*, Tech. Report 99-03, Data Mining Institute, Computer Sciences Department, University of Wisconsin, Madison, Wisconsin, September 1999, ftp://ftp.cs.wisc.edu/pub/dmi/tech-reports/99-03.ps.

[Man65] O. L. Mangasarian, *Linear and nonlinear separation of patterns by linear programming*, Operations Research **13** (1965), 444–452.

[Man68] O. L. Mangasarian, *Multi-surface method of pattern separation*, IEEE Transactions on Information Theory **IT-14** (1968), 801–807.

[Man97] O. L. Mangasarian, *Mathematical programming in data mining*, Data Mining and Knowledge Discovery **1** (1997), no. 2, 183–201, ftp://ftp.cs.wisc.edu/math-prog/tech-reports/96-05.ps.

[Man99] O. L. Mangasarian, *Arbitrary-norm separating plane*, Operations Research Letters **24** (1999), no. 1-2, ftp://ftp.cs.wisc.edu/math-prog/tech-reports/97-07r.ps.

[Man00] O. L. Mangasarian, *Generalized support vector machines*, Advances in Large Margin Classifiers (Cambridge, MA) (A. Smola, P. Bartlett, B. Schölkopf, and D. Schuurmans, eds.), MIT Press, 2000, ftp://ftp.cs.wisc.edu/math-prog/tech-reports/98-14.ps, pp. 135–146.

[MAT92] MATLAB, *User's guide*, The MathWorks, Inc., Natick, MA 01760, 1992.

[MM79] O. L. Mangasarian and R. R. Meyer, *Nonlinear perturbation of linear programs*, SIAM Journal on Control and Optimization **17** (1979), no. 6, 745–752.

[MM99] O. L. Mangasarian and David R. Musicant, *Data discrimination via nonlinear generalized support vector machines*, Tech. Report 99-03, Computer Sciences Department, University of Wisconsin, Madison, Wisconsin, March 1999, ftp://ftp.cs.wisc.edu/math-prog/tech-reports/99-03.ps.

[MSW95] O. L. Mangasarian, W. N. Street, and W. H. Wolberg, *Breast cancer diagnosis and prognosis via linear programming*, Operations Research **43** (1995), no. 4, 570–577.

[Sto74] M. Stone, *Cross-validatory choice and assessment of statistical predictions*, Journal of the Royal Statistical Society **36** (1974), 111–147.

[Str94] W. N. Street, *Cancer diagnosis and prognosis via linear-programming-based machine learning*, Computer Sciences Department, Mathematical Programming Technical Report and Ph.D. Dissertation 94-14, University of Wisconsin, Madison, Wisconsin, August 1994, http://www.cs.wisc.edu/tech-reports/.

[Vap95] V. N. Vapnik, *The nature of statistical learning theory*, Springer, New York, 1995.

[WLM99] W. H. Wolberg, Y.-J. Lee, and O. L. Mangasarian, *WPBCC: Wisconsin Prognostic Breast Cancer Chemotherapy Database*, Computer Sciences Department, University of Wisconsin, Madison, ftp://ftp.cs.wisc.edu/math-prog/cpo-dataset/machine-learn/WPBCC/, 1999.

[WSHM95a] W. H. Wolberg, W. N. Street, D. H. Heisey, and O. L. Mangasarian, *Computer-derived nuclear features distinguish malignant from benign breast cytology*, Human Pathology **26** (1995), 792–796.

[WSHM95b] W. H. Wolberg, W. N. Street, D. N. Heisey, and O. L. Mangasarian, *Computerized breast cancer diagnosis and prognosis from fine-needle aspirates*, Archives of Surgery **130** (1995), 511–516.

[WSM93] W. H. Wolberg, W. N. Street, and O. L. Mangasarian, *Breast cytology diagnosis via digital image analysis*, Analytical and Quantitative Cytology and Histology **15** (1993), no. 6, 396–404.

[WSM97] W. H. Wolberg, W. N. Street, and O. L. Mangasarian, *Computer-derived nuclear features compared with axillary lymph node status for breast carcinoma prognosis*, Cancer Cytopathology **81** (1997), 172–179.

Computer Sciences Department, University of Wisconsin, 1210 West Dayton St. Madison, WI 53706

*E-mail address*: yuh-jye@cs.wisc.edu

Computer Sciences Department, University of Wisconsin, 1210 West Dayton St. Madison, WI 53706

*E-mail address*: olvi@cs.wisc.edu

General Surgery, University of Wisconsin, H4/750 Clinical Sciences Center Madison, WI 53792

*E-mail address*: wolberg@surgery.wisc.edu

DIMACS Series in Discrete Mathematics
and Theoretical Computer Science
Volume **55**, 2000

# A Model for Learning Human Vascular Anatomy

Peter Hall

ABSTRACT. This paper gives a complete description of a representation for vascular anatomy that enables inter-patient variations to be incrementally acquired. We separate the vasculature into its topology (branching structure) and its geometry (vessel shape). For each part we provide a representation, suggest similarity measures, and define addition operations; thus allowing complete vascular representations to be incrementally acquired. We review the acquisition process, and outline a typical use of the representation: three-dimensional reconstruction of vasculature from biplanar angiograms. The novelty in this paper is furnished by the representation of inter-individual variations in vessel shape that permits incremental acquisition.

## 1. Introduction

Representations (models) of anatomy are important in medical image analysis. This is not surprising, when we consider that it takes humans years of training to interpret medical images. Such training suggests that humans must learn about anatomy, and how it appears in images of different kinds, which implies that our general purpose vision is insufficient for medical image analysis. In the context of computerised analysis general questions are how to acquire, use, and represent anatomy and its variations. Representation is a key issue, for it influences both acquisition and use. We will argue in favour of representations that can be incrementally acquired, from images say. Furthermore, because we model physical vasculature *per se*, the representation may be used in many different tasks.

Representing vascular anatomy is a challenging problem for several reasons, not least of which is the wide inter-individual variation in branching structure (topology) and vessel shape (geometry) [**SH76**]; Figure 1 provides an example. As a broad trend, larger vessels tend to show less inter-individual variation than the smaller vessels. There is sufficient regularity for clinicians to be able to name vessels, especially the larger ones. Clinical nomenclature of a vessel only weakly depends on its geometry and location within the branching pattern, instead nomenclature relies on the territories (of the brain or heart, say) that the vessel is associated with. Vessels are partitioned into the arterial system (which feed), and the venous system (which drain) territories. In general, vascular systems form a graph-like structure: vessels part in the arterial system and recombine in the venous system. The arterial system itself embodies graph-like components such as the circle of Willis [**SH76**]

2000 *Mathematics Subject Classification.* Primary 68T05, 68T45; Secondary 15A18, 62H99, 68R10.

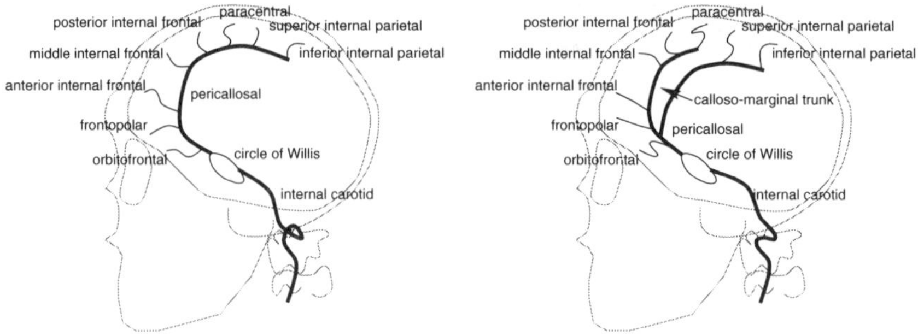


FIGURE 1. Selected variations of branches of the pericallosal artery, and its feeding vessel, the internal carotid.

found in the brain, as seen in Figure 1 and also in Figure 2. The presence of some vessels may exclude others — plausibly, because any given territory needs to be served just once — thus the shape of any given vessel may affect branching structure elsewhere in the vasculature. The three-dimensional trajectory of vessels that match between individuals will vary, usually moderately but widely on occasion. Finally, only the most common of the variations in vasculature are documented either in the clinical texts (such as [**SH76**]) or elsewhere, to the best of our knowledge. Thus a representation of vasculature that incrementally acquires such variations is advantageous.

Vascular systems can be imaged in several ways, we give three examples. When injected with an absorbent dye they can be photographed using x-rays, in which case they appear under perspective projection; this is x-ray angiography. They can also be imaged as magnetic resonance angiograms (MRA), in which case they appear sliced in a stack of images. Ultrasound is an imaging modality, which shows cross sections of the body, and from which information about the interior surface of blood vessels can be gained. Each imaging modality has a distinct set of characteristics, which should be taken in account by any automatic system for image interpretation. Furthermore, the reasons for acquiring the image is variable. For example, x-ray angiograms of the coronary vessels may be acquired to assess blood flow, while cerebral angiograms may be taken to locate any *arterio-venous malformation* in the vessel structure. Consequently, there seems to be good reason to represent vasculature in a manner that depends not only on image type, but also on task. Yet we argue against this, and in favour of representing physical vasculature in three-dimensions (and in time, if necessary). This provides the benefit of task independence, at the cost that particular tasks may be more awkward. We argue that the vascular representation can be augmented with image and task specific data, or else such information be encoded in the algorithms, thus mitigating against this affect.

Our vascular representation, briefly, is a labelled graph. The structure of the graph encodes vascular topology, while the labels encode three-dimensional geometry. Similar representations are used by others, and we now outline salient examples. Wahle *et. al.* [**WWM**$^{+}$**95**] use a three-dimensional model with a tree-like structure, which is acquired via manual interpretation of x-rays angiograms and used by clinicians for treatment planning. Similar models are used by Fessler Macovski [**FM91**] and Székely *et. al.* [**SKKG95**] to segment MRA. Garreau *et. al.* [**GCCC91**] work with x-ray angiograms and also use a three-dimensional representation, but they encode branching patterns with PROLOG rules. In the last three cases the representation is hard-coded by the programmer and cannot be updated. Furthermore, none of the above allow for variations in topology other than by provision of an (in-exhaustive) list, and none represent geometric variance.

All the above are models of physical vasculature, and so are task independent to some extent. In contrast, earlier research tended to provide representations of vasculature premised explicitly on a particular imaging modality. For example, a fixed set rules that not only explicitly describe vasculature as seen in x-ray angiograms, but also provide references to code elements needed to interpret the images [**DSSM90, RS87, SvCF$^+$87**]. The common aim of these approaches was to assign a correct clinical name to each vessel. However, the approach is not only specific to x-rays, but to x-rays taken from particular angles. Moreover, the representation works for coronary vasculature (around the heart), but tends to fail for cerebral vasculatures (in the brain) for several reasons: (1) cerebral vessels are smaller, more numerous, and more tortuous in geometry; (2) cerebral vasculatures exhibit wider inter-individual variations of topology than coronary vasculature; (3) the extent of these variations is not fully documented, either in the clinical literature or elsewhere, so a fixed rule base is sure to be incomplete.

All the above are examples use what might be called "strong knowledge", that is an explicit representation either about the vasculature itself, or about how it appears in images. However, representations exist that use "weak knowledge", by which we mean general physical models such as the fact that the connectivity of any vascular branch is invariant with respect to point of view. Sun and Bergson [**SB88**] use this to match vasculature between x-ray angiograms — the number of possible matches between one vessel in an image and many vessels in the other is very much reduced. Unfortunately, this approach can be defeated by quite simple geometric configurations, such as a tuning fork. Furthermore, the method is of no use when segmenting MRA, say. Spatio-temporal coherence of vasculature is another form of weak knowledge, and is used by Nguyen and Sklansky [**NS94**] to reconstruct vasculature from single-view cineangiograms (x-ray movies) by tracking coronary vessels from frame to frame. This form of weak knowledge can be used in other image modalities, but requires a vascular model to be set in space for the first frame.

From the above discussion, we argue for a representation of vasculature of the "strong" variety, that:

- accounts for variations in topology (branching structure);
- accounts for variations in three-dimensional geometry (vessel shape);
- can be incrementally acquired to compensate for the lack of a documented set that characterises such variations;
- and can be used in a broad variety of tasks and with a broad variety of input data types.

As mentioned above, our approach uses a labelled graph that separates vascular topology from vessel geometry. Incremental acquisition is facilitated by two operations:

- An operator that measures the similarity between two vasculature.
- An operator that adds two vasculatures.

Thus we can add a pair of vasculature, if they differ sufficiently. In the remainder of this paper we first overview our representation using an example (Section 2), then we give a general account of the topological aspects of the representation (Section 3), before discussing geometry in Section 4. In Section 5 we outline acquisition and a particular use, namely three-dimensional reconstruction, and conclude with critical remarks in Section 6.

## 2. An overview

We seek to represent a collection of physical vasculature built from individual instances; otherwise said, we wish to build a "vascular catalogue". We begin by partitioning each individual vasculature into "furcations" and "vessel fragments". A furcation is a place in the vasculature where a vessel divides into two or more (as in arteries), or else where two or more vessels recombine (as in veins). We must also recognise places on the vasculature that delimit the model, which for convenience are regarded as furcations. A vessel fragment is a piece of vasculature that connects a pair of furcations. Here, we use the term vessel fragment because *vessel*, as recognised in clinical texts [**SH76**], would correspond to a sequence of vessel fragments and furcations.

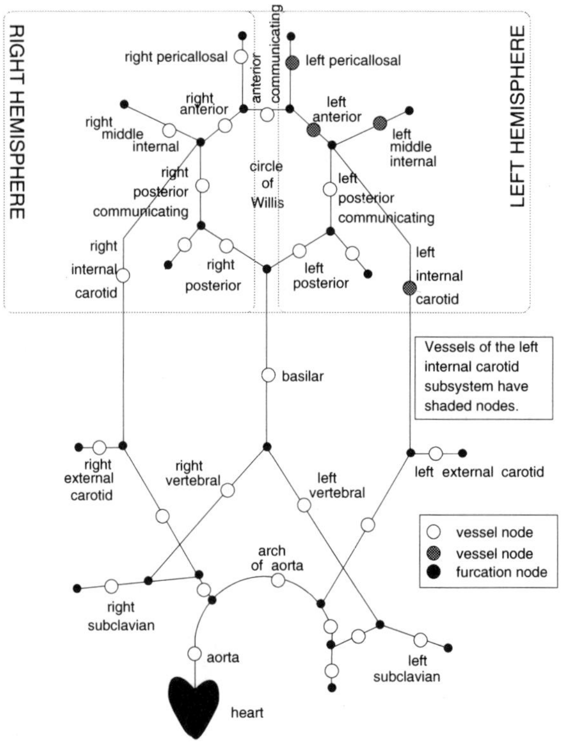


FIGURE 2. A topological map of the major arteries that arise from the heart and feed the brain. The circle of Willis, at the base of the brain, shows the graph-like structure of the cerebral arterial system.

An obvious way to map the vasculature of any individual onto a graph is to map furcations to nodes, and vessel fragments to edges. Unfortunately this mapping leads to problems when representing a collection of vasculature. One such problem is exposed when trying to combine a pair of vessel fragments that correspond in structural terms (they are in the same place in the vascular structure), but which differ in geometry to such an extent that the vessel fragments are best regarded as instances of different classes. We are faced with a conflict: topologically the vessel fragments match, geometrically they do not. The problem is how to resolve this conflict within the representation of vasculature. To make matters worse, the territories of the brain covered by a vessel fragment depend on its shape, and hence the shape of a vessel will, in general, affect the topology of other parts of the vascular structure. Again, the representation must account for this.

One approach is to use multi-edge graphs, but at the cost of excluding much of standard graph theory. Multi-labelled graphs are another alternative, in which several copies of label sets are applied to the same edge. However, the fact that geometry can affect topology, leads to difficulties. Both of these alternatives are best regarded as *ad hoc* patches, both make implementation difficult and are best avoided.

Fortunately a simple change to the mapping overcomes all these objections, and leads to a final representation for a collection of vasculature that is more compact than would otherwise be the case. The new mapping maps both furcations and vessel fragments onto graph nodes, thus we have "furcation nodes" and "vessel nodes". The important restriction is that edges of the graph can only connect furcation nodes to vessel nodes, and nodes of the same type cannot be connected. Such a graph is a *bipartite graph.* An example can be seen in Figure 2, which shows a topological view of the major arteries that arise from the heart and feed the brain. Major subsystems, such as that fed by the internal carotid artery can be identified as subgraphs. We demonstrate problems of the simple mapping, and the use of bipartite graphs, with the aid of an example.

**2.1. An example.** The problems associated with adding a pair of graphs, and their resolution using a bipartite graph are illustrated in Figure 3. We now explain this example.

Consider a pair of graphs:

$$G_1 = (\{a{:}X,\ b{:}Y,\ c{:}Z\}, \{ab{:}Q,\ bc{:}P\})$$

and

$$G_2 = (\{a{:}W,\ b{:}Y,\ c{:}Z\}, \{ab{:}Q,\ bc{:}P\})$$

in which identifiers, such as $a$, precede their label, such as $X$. (We use text labels for demonstrative convenience.) We wish to "add" these graphs. We proceed by finding their maximum common subgraph, while taking labels in account, indeed one practical affect of labelling is to make this operation computationally feasible viable. We rule that all identifiers are treated as dummy variables. Nodes and edges with matching labels can correspond only if their labels match. Furthermore, edges can correspond only if their delimiting nodes correspond, this ensures that each edge has exactly one label.

The result of addition will be another graph, in which the original pair are subgraphs. We wish to identify these subgraphs, and so augment their nodes and

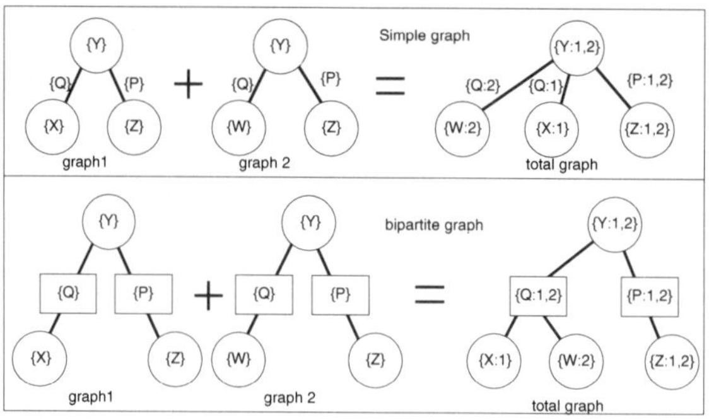


FIGURE 3. A pair of graphs with simple mapping are added in the top row. These are represented and added in bipartite form in the bottom row. The bipartite form ensures that all labels are stored exactly once. This illustration omits identifiers on nodes to emphasise that they are dummy variables.

edges with "structural labels", such as 1 or 1, 2. In the example the result is

$$G = (\{a{:}W{:}2,\ d{:}X{:}1,\ b{:}Y{:}1{,}2,\ c{:}Z{:}1{,}2\ \}, \{ab{:}Q{:}2,\ bc{:}P{:}1{,}2,\ bd{:}Q{:}1\ \})$$

Although this representation does capture both of the initial graphs as a subgraph, we see that differing labels ($\{X\}$ and $\{W\}$) can cause topological changes that are quite remote from their point of origin; on larger examples the affect is amplified. Additionally, we see a practical problem: the need to store labels, such as $Q$, more than once.

We now represent $G_1$ and $G_2$ in bipartite form, to obtain the bipartite equivalents, $H_1$ and $H_2$:

$$H_1 = (\{a{:}X,\ b{:}Y,\ c{:}Z,\ ab{:}Q,\ bc{:}P\ \}, \{(a,\ ab), (ab,\ b), (b,\ bc), (bc,\ c)\})$$

and

$$H_2 = (\{a{:}W,\ b{:}Y,\ c{:}Z,\ ab{:}Q,\ bc{:}P\ \}, \{(a,\ ab), (ab,\ b), (b\ ,bc), (bc,\ c)\})$$

Here vessel nodes and furcation modes can be discriminated by the length of their identifier. (A more general approach is to use an explicit label.) These new graphs can now be combined, again via a maximum common labelled subgraph:

$$\begin{aligned} H &= (\{a{:}W{:}2,\ d{:}X{:}1,\ b{:}Y{:}1{,}2,\ c{:}Z{:}1{,}2,\ ab{:}Q{:}1{,}2,\ bc{:}P{:}1{,}2\ \}, \\ &\quad \{(a,\ ab){:}2,\ (d,\ ab){:}1,\ (ab,\ b){:}1{,}2,\ (bc,\ b){:}1{,}2,\ (bc,\ c){:}1{,}2\}) \end{aligned}$$

This representation, which have not seen reported elsewhere, allows edges to merge regardless of the properties of nodes that delimit them. Similarly, nodes can now be merged regardless of the edges they delimit. All the properties of standard graphs are maintained (one edge between two nodes, for example) and structural labelling is consistent. Notice that the problems associated with wide variations in the geometry of vessel fragments are addressed, because each distinct geometrical feature is a distinct node. The changes in topology are more localised than in the above case, and each label is stored exactly once.

Rather than use structural labels, we can identify the component graphs by using a binary matrix. The rows of binary matrix are read to give those nodes and edges in a particular component subgraph. The columns are read to give those component subgraphs to which a given node or edge belongs. The binary matrix for the example is

| | $a$ | $b$ | $c$ | $d$ | $ab$ | $bc$ | $(a,ab)$ | $(d,ab)$ | $(ab,b)$ | $(b,bc)$ | $(bc,c)$ |
|---|---|---|---|---|---|---|---|---|---|---|---|
| $H_1$ | 0 | 1 | 1 | 1 | 1 | 1 | 0 | 1 | 1 | 1 | 1 |
| $H_2$ | 1 | 1 | 1 | 0 | 1 | 1 | 1 | 0 | 1 | 1 | 1 |

This matrix enhances task independence because it does make reading rows or columns equally easy. We now turn our attention to a more general solution.

## 3. Representing discrete variance: topology

The aim is to combine $K$ graphs, $G_i$, in a set (library) $L$, to make a compact vascular catalogue. We have

$$L = \{G_i : i = 1 \ldots K\} \tag{1}$$

$$G_i = (N_i, E_i) \tag{2}$$

where $N_i$ is a set of nodes and $E_i$ a set of edges. Each of the $G_i$ represents the vasculature of an individual, represented as a bipartite graph. However, we will not assume bipartite form in our development, but will *in situ* note the affect of a bipartite graph. This leads us to a more general solution that does not depend in any way on the specific problem of modelling a set of vasculature.

Geometry (or, indeed, any other property) is specified with a set of labelling functions $f(.)$ whose arguments we allow to vary in type. This "overloading" of $f(.)$ is intended to aid clarity by avoiding many symbols or subscripts — the intended action of $f$ is clear from its argument. We write

$$f(G_i) = f(N_i) \cup f(E_i) \tag{3}$$

$$f(n \in N_i) = \mathbf{x}_{ni} \tag{4}$$

$$f(e \in E_i) = \mathbf{y}_{ni} \tag{5}$$

Hence $f(G_i)$ is a set of geometrical descriptions. The labels, $\mathbf{x}$ for nodes and $\mathbf{y}$ for edges, are subscripted to show that each node and each edge in each graph can have a unique label. Here, they are descriptions of geometry such as those we discuss in Section 4. We note that $\mathbf{x}$ has two forms in a bipartite graph: it can represent either furcations or vessel fragments. Furthermore, $\mathbf{y}$ is a null label in the bipartite graph.

We define another labelling function, $s(.)$, to provide structural information. Given an argument, this function will identify those members of the library, $L$, to which the argument belongs. Thus we have:

$$s(G_i) = < s(N_i), s(E_i) > \tag{6}$$

$$s(n \in N_i) = \{i\} \tag{7}$$

$$s(e \in E_i) = \{i\} \tag{8}$$

Here $s(G_i)$ is an n-tuple of length $|G_i| = |N_i| + |E_i|$, with every component the set $\{i\}$. As before, we overload the function to assist clarity. Edge labels are redundant for bipartite graphs.

We define addition for a pair of such individuals to produce a *protograph.* A protograph has structural labels so that the elementary graphs from which it was made can be identified. Other members of the library can then be added to this protograph. The above labelling defines each element of $L$ as a degenerate protograph, and so provides for a consistent representation between a collection of vasculature and an individual. Hence we need only consider adding the individuals, $G_i$ and $G_j$, in the library into a single, compact object. The use of the maximal common labelled subgraph requires the need for a definition of similarity. Hence we first discuss similarity measures, and then proceed to addition.

**3.1. Similarity.** We define the maximum common labelled subgraph of $G_i$ and $G_j$, as follows: Let $g_i = (M_i, A_i)$ be a labelled subgraph of $G_i$, and $g_j = (M_j, A_j)$ be a subgraph of $G_j$, with $M \subseteq N$ and $A \subseteq E$ in each case. Let $g_i$ be isomorphic to $g_j$, by which we mean there is a bijection, $h : M_i \mapsto M_j$, and two functions $D_N$ and $D_E$, such that:

$$(m, n) \in A_i \Leftrightarrow (h(m), h(n)) \in A_j \tag{9}$$

$$D_N(f(n), f(h(n))) \leq \epsilon_N \tag{10}$$

$$D_N(f(m), f(h(m))) \leq \epsilon_N \tag{11}$$

$$D_E(f(m, n), f(h(m), h(n))) \leq \epsilon_E \tag{12}$$

all hold, as do equivalent expressions when the inverse function $h^{-1}$ is used to map $M_j$ to $M_i$. The distance measures must be symmetric in the sense that $D(a, b) = D(b, a)$. Then $g_i$ (which is equivalent to $g_j$) is a common subgraph of both $G_i$ and $G_j$. If the number of nodes and edges in $g_i$ is maximal, then the subgraph is a *maximal common labelled subgraph.*

In the above, Equation 9 asserts that edges between corresponding nodes should also correspond. Equations 10 and 11 assert that the distance between the geometric descriptions of corresponding nodes be pleasingly small (less than some $\epsilon$), and Equation 12 asserts the same for edges. At this stage, for the sake of generality, we chose not to be prescriptive about the form of the $D$ measures; we have already specified a necessary property, and will make the model specific by suggesting particular $D$ in Section 4. Note that for a bipartite graph in that the $D_N$ has up to two forms (one for each kind of node), and the $D_E$ is trivially satisfied.

We will say that two nodes are equivalent, if they correspond according to the above definition. Edge equivalence is defined in an analogous manner. Thus the maximum common labelled subgraph is the largest intersection between the sets, $f(G_i)$ and $f(G_j)$, subject to topological constraints provided by the graphs $G_i$ and $G_j$. We write

$$g = G_i \otimes G_j \tag{13}$$

for the maximal common labelled subgraph between two graphs.

We can approach a definition of similarity between graphs $G_i$ and $G_j$ using the maximal common subgraph. We observe that there will be some nodes and edges (features) in each graph that are not in $G_i \otimes G_j$. These features define the symmetric difference $\tilde{g}$ (which can be partitioned into two disconnected parts using topological constraints in $G_i$ and $G_j$). Further, the union set of features can be defined as $g \cup \tilde{g}$, again subject to topological constraints, and written as

$$G = G_i \oplus G_j \tag{14}$$

Hence, one measure of the similarity between a pair of graphs is

$$\frac{|G_i \otimes G_j|}{|G_i \oplus G_j|} \tag{15}$$

in which $|G| = |N| + |E|$. Other definitions may be preferred for specific tasks. Indeed, we use an alternative measure when finding corresponding vessels in a pair of x-ray angiograms [**HNA97**].

**3.2. Addition.** Addition of two graphs $G_i$ and $G_j$ makes use of the above definitions, most explicitly the notion of their union $G_i \oplus G_j$. Hence we can add a pair of graphs, as follows:

$$(16) \quad G = G_i \oplus G_j = (N_i, E_i) \oplus (N_j, E_j) = (N_i \oplus N_j, E_i \oplus E_j) = (N, E)$$

Clearly the addition operator is distributive over nodes and edges. In the simple case in which node/edge identifiers double as labels (an "eponymous" graph) the operator is set union. Here the labels are added as follows, under the supposition that nodes $n_i \in N_i$ and $n_j \in N_j$ correspond to give node $n \in N$, similarly for edges:

$$(17) \quad s(n) = s(n_i) \oplus s(n_j) = s(n_i) \cup s(n_j)$$

$$(18) \quad s(e) = s(e_i) \oplus s(e_j) = s(e_i) \cup s(e_j)$$

$$(19) \quad f(n) = f(n_i) \oplus F_N(n_j) = F_N(f(n_i, n_j))$$

$$(20) \quad f(e) = f(e_i) \oplus F_N(e_j) = F_E(f(e_i, e_j))$$

Structural labels are simply subject to set union. The addition of geometry is complicated by the fact that we would like to add shapes that are "close", according the measure $D$. Therefore, a simple union is not sufficient. Hence we require functions $F$ to combine the input labels to produce a new label of the same form.

There is the implicit understanding that $s(n \in N_i)$ and $f(n \in N_i)$ both yield invalid values if $n \notin N_i$. Edges are treated in analogous fashion.

Again, edge labelling is redundant for graphs in bipartite form, so that $F_E$ is trivial in that case, while $F_N$ has up to two forms.

The addition operation between a pair of graphs thus yields a new graph which can then participate in more pair-wise additions. This enables many graphs to be added. In particular, those in the library, $L$ may be added:

$$(21) \quad G = \bigoplus_{i=1}^{K} G_i = \left( \bigoplus_{i=1}^{K-1} G_i \right) \oplus G_K$$

Thus we can incrementally generate a protograph, $G$. The data for any increment can come from an arbitrary source, though we are thinking of using reconstructions from images.

Recalling the example in Subsection 2.1, a useful refinement is made by storing structural information in a binary matrix. Adding a new graph into the protograph will require the matrix to be updated only if that new graph is not currently a component subgraph; in such a case new columns must be added for each new physical feature (whether it maps to a node or an edge), and a whole new row to specify a new component subgraph. The set-like operations, such as $\otimes$, specify the new columns and rows. The construction of a protograph and binary matrix is illustrated in Figure 4, which represents a real section of cerebral vasculature; only topology is represented.

Thus far we have added graphs in such a way that individual components can be retrieved, and hence shown that topological variance can be represented and acquired incrementally. However, new instances not in the given library can be generated by "stitching together" pieces of graph from different individuals, so we make construction such as "this vasculature is made of parts of person A and

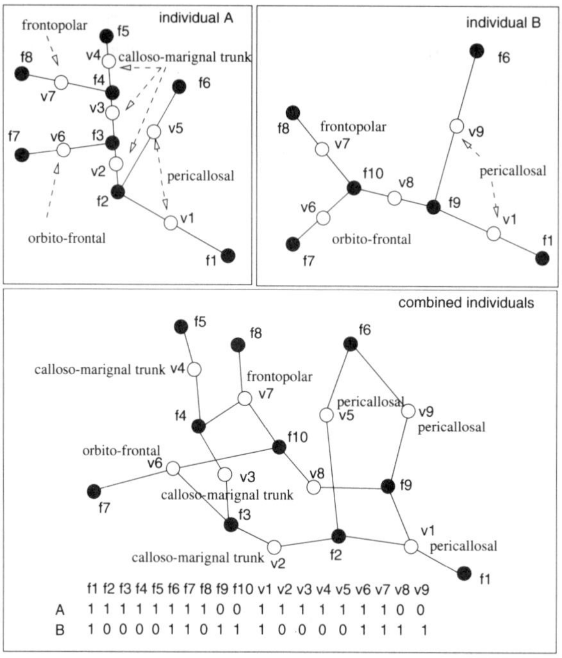


FIGURE 4. A pair of topological variations can be represented as subgraphs of a single graph (a protograph) that combines them. The individual subgraphs can be identified by using a binary matrix that specifies which vasculature and features belong together. Text labels are for illustration only.

parts of person B". This is useful because it enables our reconstruction algorithm to speculate novel structures by combining pieces from different vasculature, as outlined in Section 5.

We have left open the specification of the labels $f$, and the definition of distance measures $D$ and addition functions $F$, which act on these labels. This means that the presentation thus far is quite general, and could apply to structures quite different from vasculature. We now make it specific to vasculature by defining the geometric labels and their associated functions.

## 4. Representing continuous variance: geometry

We need to represent geometry in a way that accounts for inter-individual variance, and is incrementally updatable. Furthermore, a representation that is invariant with respect to scale, orientation, and location is desirable. We would also like to represent geometry in a way that is task independent. Unfortunately task independence is not wholly possible, nevertheless we can use a representation that admits a wide range of tasks. Where we are forced into a choice, we prefer those representations that favour automated image interpretation. In any case, we argue that augmenting the labels of the protograph can provide task dependent information.

We choose eigenspace models to represent variations in geometry, as we explain below. This method has many advantages, one of which is to unite the geometric features that would otherwise divide into two broad classes; vessel fragments and furcations. Thus we need only define one similarity measure, and one addition function. Furthermore, eigenspace models account for variations in geometry in a statistically acceptable way, and are useful for image processing tasks; the point distribution models (PDMs) of Cootes *et. al.* [**CTCG92**] are an example. Importantly, we have recently provided a method for adding eigenspaces [**HMM**], which makes their incremental acquisition a feasible proposition. However, we begin by explaining how we represent a single instance of a furcation, and a single instance of a vessel fragment, both seen in Figure 5.

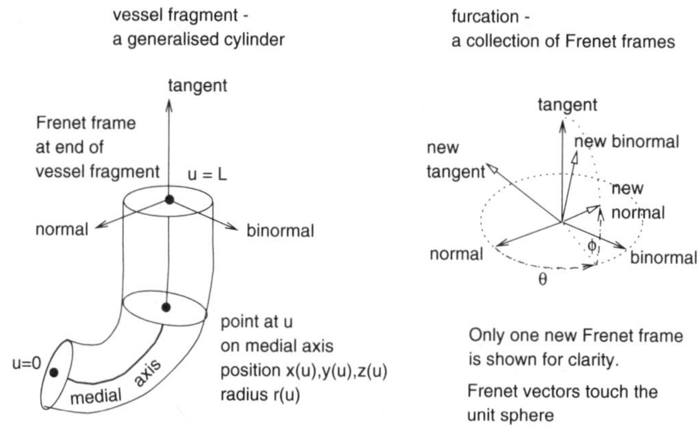


FIGURE 5. A diagram showing the geometry of a vessel fragment and a furcation. Frenet frames align furcations and vessel fragments.

We represent a single vessel fragment with a generalised cylinder $\mathbf{Q}(u) = ((x(u), y(u), z(u)), r(u), L)$ in which $u$ is arc length along the curve, which is of total length $L$. $(x(u), y(u), z(u))$ is the spatial location at $u$, and $r(u)$ specifies the normal radius at $u$. We can represent the axis of a vessel fragment in many ways, examples include a series of three-dimensional points, or a piecewise cubic curve. We can transform between such representations, using whichever form is most convenient at use time and hence need not be over prescriptive. What is important here is that we can always write the description of a vessel fragment as a column vector, which labels vessel nodes. Such column vectors are data in the eigenspace model of geometric variation for vessel fragments in the same class.

We represent the geometry of a single furcation via Frenet frames [**DS79**], which are orthonormal basis sets, as illustrated in Figure 5. A Frenet frame can be associated with each point, at $u$, on the medial axis of a vessel segment. At a furcation three or more Frenet frames each share a common origin (unless the furcation delimits the model, in which case there is only one Frenet frame). We can make the description independent of absolute orientation by specifying a "base frame" to be the frame of the vessel nearest to the heart (the single feeding vessel in the arterial system, the single draining vessel in the venous system). Other frames can be specified relative to this with two rotations about the base frame: first around the base tangent, and then around the newly rotated binormal. A pair of such rotations is needed for every Frenet frame at the furcation, except the base frame. The description is stored as a column vector, and furcation node. Such a vector is a datum in the eigenspace model of geometric variation for furcations in the same class.

Thus both furcations and vessel fragments are represented as column vectors. In general, we write as $\mathbf{x}$. The size of this vector may vary between classes. For example the number of points along the medial axis of a vessel fragment, or the number of vectors in a furcation. Even so, the discussion now uses a vector to refer to both furcations and vessel fragments.

Our objective is to represent a collection of $N$ column vectors $\mathbf{x}_i$ in eigenspace form, each $\mathbf{x} \in \Re^n$ is $n$-dimensional. The collection is represented by an $(N \times n)$ matrix, $X_{nN}$, we use subscripts to indicate matrix size. We assume that each vector corresponds in that it represents the same geometrical object, and that each vector is aligned as closely as possible with some reference object. We further assume that the $\mathbf{x}_i$ are Gaussian distributed around their arithmetic mean. We compute the mean and covariance of $X_{nN}$:

$$\mu(X) = \frac{1}{N}\sum_{i=1} \mathbf{x}_i \tag{22}$$

$$C(X)_{nn} = \frac{1}{N}\sum_{i=1} N(\mathbf{x}_i - \mu(X))(\mathbf{x}_i - \mu(X))^T \tag{23}$$

We compute the eigenvalue decomposition of $C_{nn}$, and approximate it with the first $p$ eigenvectors and eigenvalues:

$$C(X)_{nn} = U(X)_{nn}\Lambda(X)_{nn}U(X)_{nn}^T \tag{24}$$

$$\approx U(X)_{np}\Lambda(X)_{pp}U(X)_{np}^T \tag{25}$$

in which $U$ is a matrix whose columns are eigenvectors, and $\Lambda$ is a diagonal matrix. The eigenvectors are an orthonormal basis. The error is $U(X)_{nd}\Lambda(X)_{dd}U(X)_{nd}^T$, which are the discarded eigenvectors and eigenvalues. The kept eigenvectors describe a subspace of $\Re^n$, centred at $\mu(X)$. The eigenvalues indicate the spread of the data $\mathbf{x}_i$ through this subspace: specifically, they are a measure of standard deviation. We specify an eigenspace as

$$\Omega(X) = (\mu(X), U(X)_{np}, \Lambda(X)_p, N(X)) \tag{26}$$

and visualise it as a $p$-dimensional hyperellipsoid in $n$-dimensional space. Typically $p < n$, whence $\Omega(X)$ is a compression of the data. Note we keep the number of vectors $N(X)$, this is used when adding (or subtracting) a pair of eigenspaces.

New instance vectors (that is, new furcations and vessel fragments) can be generated by linear combination of the eigenvectors and the mean, subject to the constraints imposed by the eigenmodel:

$$\mathbf{x}_{new} = \mu(X) + \sum_{i=1}^{p} a_i U_i \tag{27}$$

where the $a_i$ are scale factors, typically restricted to lie within three standard deviations, that is $|a_i| \leq 3(\lambda_i)^{1/2}$. We now discuss similarity and addition.

**4.1. Similarity.** A single, definitive, measure of similarity cannot be given because any particular measure for $D$ is application dependent. Furthermore, the issue is complicated because we may need to compute the similarity between a pair of single instance vessels, or between an instance and a collection of vessels, or between a pair of collections. We propose to use a measure based on difference between eigenspaces, and do not consider geometry. This has the advantage of generality, but the disadvantage that is may not be specific enough to highlight aspects that may be important to some applications. However, we emphasise that this measure is not yet fully tested and hence is speculative in nature.

Given eigenspace models $\Omega(X)$ and $\Omega(Y)$, we compute an orthonormal basis set $\nu_{nt}$ that spans the vectors $[U(X)_{np}, U(Y)_{nq}, \mu(X) - \mu(Y)]$. This basis is the smallest subspace of $\Re^n$ that spans both eigenspaces and the space between them, and has $t \leq p + q + 1$ vector components. The difference between the covariance matrices in this space is:

$$d(\Omega(X), \Omega(Y)) = abs(det(E_{tt})) \tag{28}$$

in which

$$E_{tt} = N(X)(A_{tp}\Lambda_{pp}A_{tp}^T + a_{t1}a_{t1}^T) + N(Y)(B_{tq}\Lambda_{pp}B_{tq}^T + b_{t1}b_{t1}^T) \tag{29}$$

$$A_{tp} = \nu_{nt}^T U(X)_{np} \tag{30}$$

$$a_{t1} = \nu_{nt}\mu(X) \tag{31}$$

$$B_{tq} = \nu_{nt}^T U(Y)_{nq} \tag{32}$$

$$b_{t1} = \nu n t^T \mu(Y) \tag{33}$$

This is in fact a measure of *dissimilarity* so that $d(\Omega(X), \Omega(X)) = 0$. The measure is symmetric, $d(\Omega(X), \Omega(Y)) = d(\Omega(Y), \Omega(X))$. The triangular inequality is not obeyed, so the measure is not a metric. However, it rises linearly with the distance between the means, polynomially with respect to scale of eigenvalues, and cyclically with respect to orientation of eigenvectors. It can measure the distance between eigenspace models with different dimensions (different number of eigenvectors), and handles all the combinations of single instance and collection of instances in a uniform way. Finally, it computes in a space of minimum size.

Our opinion is that the measure is worthy of further consideration, because of its useful properties, and despite of its not being a metric nor yet fully tested.

**4.2. Addition.** Suppose that we have a pair of eigenspaces $\Omega(X)$ and $\Omega(Y)$, built from data sets $X$ and $Y$ respectively. The vectors in these data sets must be of the same dimension (and in practice are supposed to be distinct instances of the same geometrical feature). We wish to add these eigenspaces, with a function $F$, to produce the eigenspace $\Omega(Z)$ for the concatenated data sets $Z = [X, Y]$. The new mean and number are easy to compute

$$N(Z) = N(X) + N(Y) \tag{34}$$

$$\mu(Z) = (N(X)\mu(X) + N(Y)\mu(Y))/N(Z) \tag{35}$$

To compute the new eigenvectors and eigenvalues we must solve the following eigenproblem

$$\begin{aligned} &\frac{N(X)}{N(Z)}\begin{bmatrix} \Lambda(X)_{pp} & 0_{pt} \\ 0_{tp} & 0_{tt} \end{bmatrix} + \\ &\frac{N(Y)}{N(Z)}\begin{bmatrix} G_{pq}\Lambda(Y)_{qq}G_{pq}^T & G_{pq}\Lambda(Y)_{qq}\Gamma_{tq}^T \\ \Gamma_{tq}\Lambda(Y)_{qq}G_{pq}^T & \Gamma_{tq}\Lambda(Y)_{qq}\Gamma_{tq}^T \end{bmatrix} + \\ &\frac{N(X)N(Y)}{N(Z)^2}\begin{bmatrix} g_p g_p^T & g_p\gamma_t^T \\ \gamma_t g_p^T & \gamma_t\gamma_t^T \end{bmatrix} = R_{ss}\Lambda(Z)_{ss}R_{ss}^T \end{aligned} \tag{36}$$

in which

$$g_p = U(X)^T(\mu(X) - \mu(Y)) \tag{37}$$

$$G_{pq} = U(X)^T U(Y) \tag{38}$$

$$H_{nq} = U(Y) - U(X)G_{pq} \tag{39}$$

$$\nu_{nt} = \mathsf{Orthobasis}(\zeta[U(X), H(Y), \mu(X) - \mu(Y)]) \tag{40}$$

$$\Gamma_{tq} = \nu_{nt}^T U(Y)_{nq} \tag{41}$$

$$\gamma_t = \nu^T(\mu(X) - \mu(Y)) \tag{42}$$

$\zeta$ is an operation that removes very small column vectors from the matrix, and Orthobasis computes a set of mutually orthogonal, unit vectors that support its argument; typically Gramm-Schmidt orthogonalisation [**GL83**] is used to compute significant support vectors. The eigenvectors we seek are given by the above computation. The new eigenvectors and eigenvalues can then be computed as

$$U(Z)_{sn} = [U(X)_{pn}, \nu_{tn}]R \tag{43}$$

This result takes into account a change of mean and differing numbers of eigenvectors in each model. It computes without reference to any original data or the original correlation matrices, and computes with minimal complexity. A single data vector, $x$ can be added into the set if its eigenmodel is defined to be $(x, 0, 0, 1)$. This procedure is fully tested; a proof and further details are available elsewhere [**HMM**].

## 5. Acquisition and use

To use the model in any way — including comparisons between models — it is a truism that models must exist. Unfortunately none exist in the first instance and so question is raised as initial acquisition: from what data source, and how? The answer is that an expert clinician builds a model using image data, such as x-ray angiograms, and a suitable user interface. This manually constructed model then becomes the initial (and at this stage, only) element in the library, and hence is a degenerate protograph. This protograph is used to assist in the automatic image interpretation of a second set of images. However, errors are likely and the expert is requested to edit and verify the automatically produced model. Once this is done the second model can be added to the first, to produce a new protograph that contains them both. The hope is that as more and more models are added into the protograph, so automatic image interpretation improves. This is why incremental acquisition may be called "learning".

The model has many use, including the simulation of x-ray angiography [**Hal98**], and the linking of medical texts with medical images [**KH96**]. However, its intended use is three-dimensional reconstruction of vasculature from x-ray angiograms, taken at about ninety degrees apart (*biplanar angiograms*), The model is necessary to help decide which vessels correspond between views; reconstruction of vessels is a separate step [**Hal97**]. Correspondence between biplanar angiograms is solved by projecting the model onto the input images, and attempting to match model vessel to the images of real vessels. The PDM [**CTCG92**] approach assists in each image because it eliminates candidate vessels that deviate too far from the average. Since PDMs are examples of eigenmodels, our eigenmodel approach fits neatly with such previous work. The fact that the model is three-dimensional means that information from both images can be used in mutual corroboration, and that occlusions of all kinds can be recognised and circumvented; thus even vessels falling entirely behind a vessel in one image (but not the other image) can be placed into correspondence. Moreover, the search for a suitable model topology is efficient and can generalise to new instances. The efficiency derives because individuals are subgraphs, hence matching a feature from the protograph is equivalent to matching a feature in several subgraphs in parallel. If the match provides only weak evidence for the vessel, then the evidence for each individual in that group can be turned down. The generalisation derives from the same source; the single protograph allows us to "mix-and-match" partial structures from distinct individuals. See [**HNA97, HNA96**] for detailed descriptions of this particular application.

## 6. Conclusion

We have presented a model that learns (incrementally acquires) inter-individual variations in topology and geometry. The protograph provides a method for compressing the topological representation of the data in $L$. The labels $s(.)$ allow us to retrieve those individuals used to make the protograph. The geometrical labels are also compressed, though statistically with the use of eigenspace models. In addition, the representation can generalise to new instances, both topologically and geometrically.

The model has many uses, but was initially designed for solving correspondence in x-ray angiograms. The learnable nature of the model was a response to the fact that no full catalogue of variations exist. The utility of the model may be improved or extended with additional labelling, provided that such labels can be incrementally acquired, and a distance measure can be defined for them.

Our geometric descriptions make use of eigenmodels. We have provided a way in which to add eigenspaces, and suggested a possible measure of dissimilarity. A criticism to be aimed here is that single eigenmodels will not capture any non-linearity on the covariance, and hence a mixture of Gaussian approach might be preferred. (Fortunately, we can still "add" such mixture, but the distance measure becomes more complex and, probably, *ad-hoc* in such circumstances.)

Despite its weaknesses the representation is quite complete and quite general; it could be used to represent much else other than vasculature. As a model for learning human vascular anatomy, the model has shown itself to be convenient and useful in several distinct scenarios.

## References

[CTCG92] T.F. Cootes, C.J. Taylor, D.H. Cooper, and J. Graham, *Training models of shape from sets of examples*, Proc. British Machine Vision Conference, 1992, pp. 9–18.

[DS79] H.F. Davis and A.D. Snider, *Introduction to vector analysis*, fourth ed., Allyn and Bacon Inc., 1979.

[DSSM90] D. Delaere, C. Smets, P. Suetens, and G. Marchal, *A knowledge-based system for the angiographic reconstruction of blood vessels from two angiographic projections*, Proc. of the North Sea Conference on Biomedical Engineering, November 1990, pp. NS27–NS36.

[FM91] J.A. Fessler and A. Macovski, *Object-based 3d reconstruction of arterial trees from magnetic resonance angiograms*, IEEE Transactions on Medical Imaging **10** (1991), no. 1, 25–39.

[GCCC91] M. Garreau, J.L. Coatrieux, R. Collerec, and C. Chardenon, *A knowledge based approach for reconstruction and labelling of vascular networks from biplane angiographic projections*, IEEE Transactions on Medical Imaging **10** (1991), no. 2, 122–131.

[GL83] G.H. Golub and C.F. Van Loan, *Matrix computations*, Johns Hopkins, 1983.

[Hal97] P. Hall, *Robust reconstruction of 3d space-curves from images at arbitrary angles*, Proc. British Machine Vision Conference, 1997, pp. 300–309.

[Hal98] P. Hall, *A complete simulation of x-ray angiography*, Proc. SPIE Medical Imaging, Volume 3338 (San Diego), 1998, pp. 1064–1075.

[HMM] P. Hall, D. Marshall, and R. Martin, *Merging and splitting eigenspaces*, IEEE Transactions on Pattern Analysis and Machine Intelligence.

[HNA96] P. Hall, M. Ngan, and P.M. Andreae, *Reconstruction of blood vessel networks from x-ray projections and a vascular catalogue*, European Conference on Computer Vision, (Lecture Notes in Computer Science 1065, Springer-Verlag) (B. Buxton and R. Cipolla, eds.), 1996, pp. 293–303.

[HNA97] P. Hall, M. Ngan, and P. Andreae, *Reconstruction of vascular networks using three-dimensional models*, IEEE Transactions on Medical Imaging **16** (1997), no. 6.

[KH96] P. Mc Kevitt and P.M. Hall, *Automatic interpretation of angiograms*, Artificial Intelligence Review **10** (1996), no. 3-4, 235–252.

[NS94] Thinh V. Nguyen and Jack Sklansky, *Reconstructing the 3-D medial axes of coronary arteries in single-view cineangiograms*, IEEE Transactions on Medical Imaging **13** (1994), no. 1, 61–73.

[RS87] S.T. Rake and L.D.R. Smith, *The interpretation of x-ray angiograms using a blackboard control architecture*, Proc. Computer Assisted Radiology (Berlin), 1987, pp. 681–685.

[SB88] Y. Sun and E. Bergerson, *Automated 3d reconstruction of tree-like structures from two orthogonal views*, International Conference on Acoustics, Speech, and Signal Processing (New York), vol. II, IEEE Press, April 1988, pp. 1296–1299.

[SH76] G. Salamon and Y. P. Huang, *A radiological anatomy of the brain*, Springer Verlag, Berlin, 1976.

[SKKG95] G. Székely, Th. Koller, R. Kikinis, and G. Gerig, *Structural description and combined 3-D display for superior analysis of cerebral vascularity from MRA*, Medical Imaging (L. Beolchi and M.H. Kuhn, eds.), Studies in Health Technology and Informatics, vol. 19, IOS Press, 1995, pp. 183–194.

[SvCF+87] P. Suetens, J. van Cleynenbreugel, F. Fierens, C. Smets, A. Oosterlink, and G. Wilms, *An expert system for blood vessel segmentation on subtraction angiograms*, SPIE Vol. 767 Medical Imaging, 1987, pp. 454–459.

[WWM+95] A. Wahle, E. Wellnhofer, I. Mugaragu, H.U. Sauer, H. Oswald, and E. Fleck, *Assessment of diffuse coronary artery disease by quantitative analysis of coronary morphology based upon 3-D reconstruction from biplane angiograms*, IEEE Transactions on Medical Imaging **14** (1995), no. 2, 230–241.

MATHEMATICAL SCIENCES, COMPUTING GROUP, BATH UNIVERSITY, ENGLAND
*E-mail address*: `maspmh@maths.bath.ac.uk`

DIMACS Series in Discrete Mathematics
and Theoretical Computer Science
Volume **55**, 2000

# Optimization of Gamma Knife Radiosurgery

Michael C. Ferris and David M. Shepard

ABSTRACT. The Gamma Knife is a highly specialized treatment unit that provides an advanced stereotactic approach to the treatment of tumors, vascular malformations, and pain disorders within the head. Inside a shielded treatment unit, beams from 201 radioactive sources are focused so that they intersect at the same location in space, resulting in a spherical region of high dose referred to as a shot of radiation. The location and width of the shots can be adjusted using focussing helmets. By properly combining a set of shots, larger treatment volumes can be successfully treated with the Gamma Knife.

The goal of this project is to automate the treatment planning process. For each patient, an optimization seeks to produce a dose distribution that conforms closely to the treatment volume. The variables in the optimization can include the number of shots of radiation along with the size, the location, and the weight assigned to each. Formulation of such problems using a variety of mathematical programming models is described, and the solution of several test and real-patient examples is demonstrated.

## 1. Introduction

The Leksell Gamma Knife is a highly specialized treatment unit that provides an advanced stereotactic approach to the treatment of tumors, vascular malformations, and pain disorders within the head [**Gan97**]. Over 100 Gamma Knife units are installed worldwide, and more than 20,000 patients are treated each year. Inside the shielded treatment unit (see Figure 1), beams from 201 Co-60 radioactive sources are focused so that they intersect at the same location in space. The result is a spherical region of high dose referred to as a shot of radiation. Four different shot sizes are available to the user. By combining multiple shots of radiation, the treatment plan can be customized to treat lesions of varying sizes and shapes.

The goal of this project is to automate the treatment planning process. For each patient, the optimization seeks to produce a dose distribution that conforms closely to the treatment volume. The variables in the optimization can include the number of shots of radiation along with the size, the location, and the weight assigned to each. This paper will address the approaches that we have investigated for solving this problem. The advantages and disadvantages of two approaches will be discussed, and optimized patient plans will be shown.

---

2000 *Mathematics Subject Classification.* Primary: 92C50; Secondary: 90C30, 90C11.

This material is based on research supported in part by National Science Foundation Grants CDA-9726385 and CCR-9972372, the Air Force Office of Scientific Research Grant F49620-98-1-0417 and Microsoft Corporation.

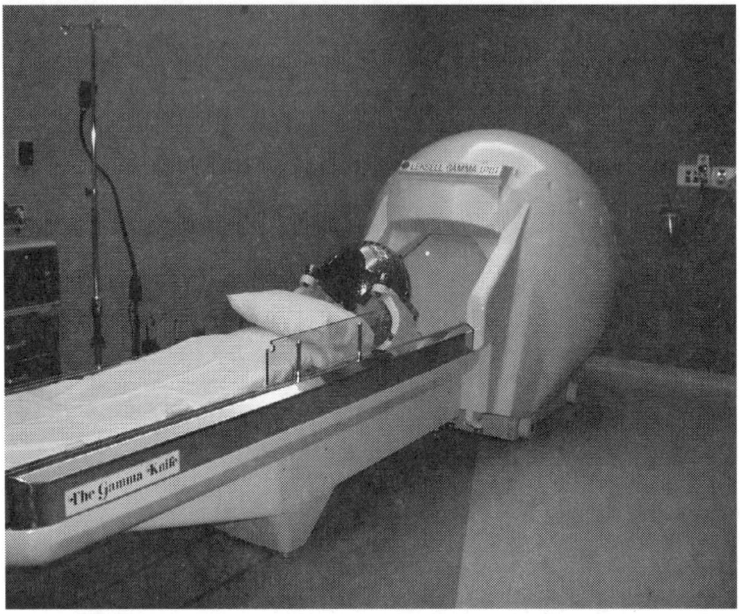


FIGURE 1. Gamma Knife Treatment Unit

The success of radiosurgery, in large part, depends upon the ability to accurately locate the treatment volume. The localization procedure begins with the fitting of a stereotactic head frame onto the patient's head (see Figure 2). This

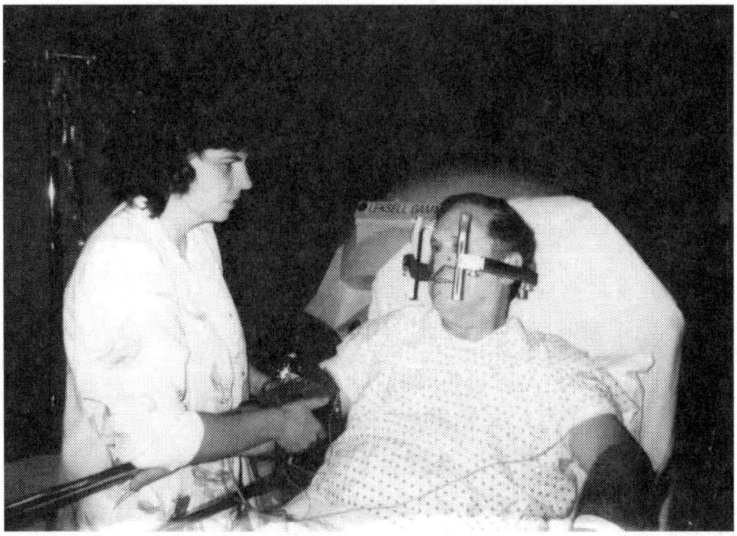

FIGURE 2. Stereotactic Head Frame

head frame is screwed into the patient's skull on the morning of the treatment.

Once the head frame is in place, the patient undergoes MRI or CT imaging. The imaging studies make it possible to precisely determine the position of the treatment volume with respect to the stereotactic head frame.

During the treatment, the patient's head frame attaches to a focusing helmet (see Figure 3). This focusing helmet serves as a collimator that further reduces the

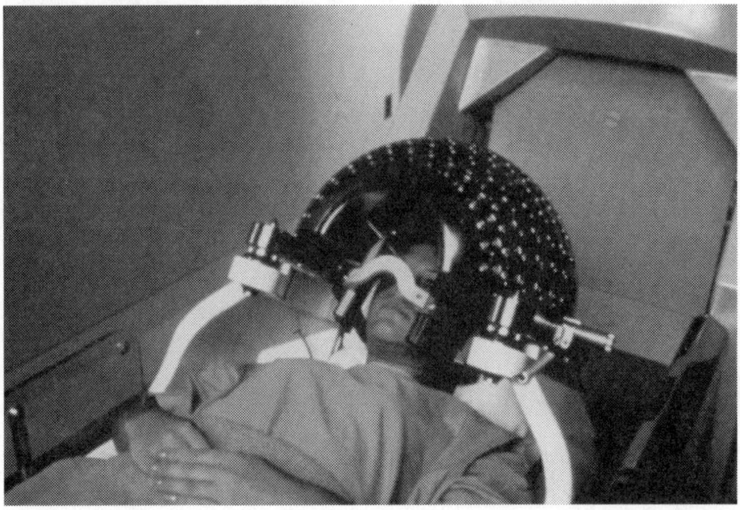

FIGURE 3. Focussing Helmet Attached to Frame

size of each beam. There are a total of four focusing helmets. These helmets can be used produce a shot of radiation that is 4mm, 8mm, 14mm, or 18mm in diameter (see Figure 4).

The neurosurgeon, the radiation oncologist, and the physicist work together in order to develop the patient's treatment plan. For some cases, the treatment planning process is relatively straightforward. For example, some small lesions can be covered with a single shot of radiation. Consequently, the treatment planning process can be completed in a few minutes. Unfortunately, the treatment planning process becomes much more complex when the tumor volume is large or irregularly shaped. These cases typically require several shots of radiation. Through an iterative process, the planners must determine the number of shots of radiation that are required along with the size, the location, and the weight that should be assigned to each. Each time that there is a change in the shot size, the staff must remove the current focusing helmet and replace it with the helmet of the correct size. This is a fairly time consuming process, because each helmet weighs approximately 500 pounds and must be manipulated mechanically. For each shot of radiation, the location of the center of the shot is determined by the connection between the helmet and the patient's head frame. Adjustments in this connection are made between each shot. At the time of delivery, the door to the treatment unit opens and the patient couch is advanced inside the shielded treatment vault (see Figure 5).

For many patients, the treatment planning process becomes both tedious and time consuming, and the quality of the treatment plan that is produced depends upon both the experience and the patience of the planner. Because of these factors, we have sought to develop an automated process for creating Gamma Knife

FIGURE 4. Stored Focussing Helmets for 4mm, 8mm, 14mm and 18mm shots

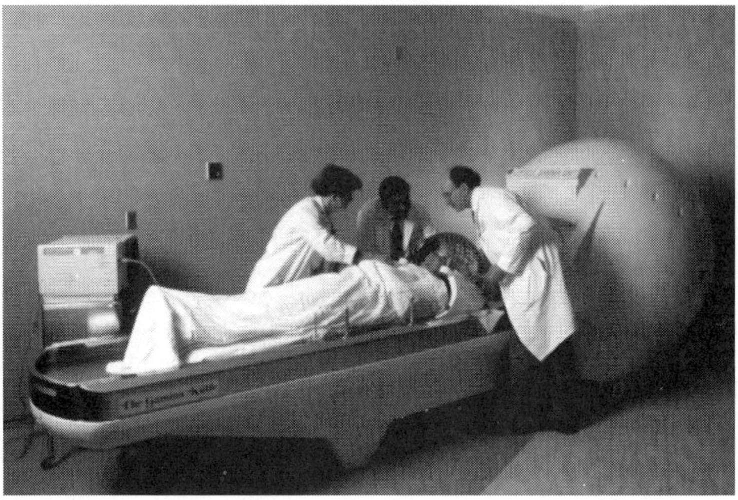

FIGURE 5. Operation of the Gamma Knife

treatment plans. As a first step in this process, the planner outlines the treatment volume and any sensitive structures. A series of treatment goals are then defined. Based upon these goals, an optimization algorithm determines the best possible treatment plan.

The ideal optimization technique for Gamma Knife treatment planning must be fast, flexible, and robust. The system must be fast, because the treatment planning process cannot begin until after the patient's head frame is in place and MRI or CT images have been obtained. It is therefore unacceptable to have a planning procedure that takes hours to complete. Our goal has been to create an automated planning system that can produce a treatment plan in 20 minutes or less.

The system must also be flexible. This is because the treatment goals will vary from one patient to the next. Finally, the system must be robust enough that it produces a high quality solution for treatment volumes of all sizes and shapes. A robust technique ensures a reasonable result regardless of the quality of the initial guess. An automated treatment planning process can potentially provide several benefits. Most importantly, the quality of patient care could be improved. Due to the large number of variables involved, it is unlikely that a manual approach will produce a truly optimal dose distribution. Improvements in the quality of a dose distribution can lead to an increase in the the probability of uncomplicated tumor control. An automated approach to treatment planning would also greatly reduce the amount of time that the neurosurgeon must commit to treatment planning. Finally, with optimization, it may be possible to reduce the number of shots and the variation in the shot size without sacrificing the quality of a patient's dose distribution. Therefore, the total treatment time would be reduced. Reduced treatment times are appreciated by the patient, and can make it possible to treat more patients in a day.

A number of researchers have studied techniques for automating the Gamma Knife treatment planning process [**WB99, LSY$^+$99**]. One approach incorporates the assumption that each shot of radiation can be modeled as a sphere. The problem is then reduced to one of geometric coverage, and a ball packing approach [**SSV93, WB99**] can be used to determine the shot locations and sizes. The use of a modified Powell's method in conjunction with simulated annealing has also been proposed [**LSY$^+$99, YSB97**].

This paper addresses a different approach, whereby the actual dose distribution is modeled and a formal constrained optimization model is solved to determine the treatment plan. In Section 2, we outline a nonparametric optimization approach for generating a dose model for the Gamma Knife. This dose model is a critical component for developing the ensuing optimization models. In Section 3, we describe two different optimization formulations that can be used for the treatment planning problem, one based on mixed integer programming, the other on nonlinear programming. We demonstrate the effectiveness of these approaches in Section 4 by comparing them on several test examples that we generate in a new testing environment. We also report computational results on several real patient cases using the nonlinear programming approach. The modeling and optimization uses various combinations of GAMS, MATLAB, CPLEX and CONOPT. The paper concludes with several remarks regarding incorporation of the methodology into the real patient planning process, along with some pointers regarding future work.

## 2. Dose Model

Each patient is currently modeled on a three dimensional grid of pixels. Given such a grid, the first task is to determine the dose that will be delivered by a particular plan at each pixel. We require an algebraic model of the distribution of the dose for use in our optimization formulations. In this model, we let $\mathcal{S}$ represent the set of the shots that we will consider, and $\mathcal{W}$ represent the possible shot sizes (typically 4mm, 8mm, 14mm and 18mm).

The complete dose distribution can be calculated as a sum of contributions from each shot delivered, once the location of the center of that shot $(x_s, y_s, z_s)$ is known, and the length of time of delivery $t_{s,w}$ is known. In practice this means that for all $(i, j, k)$

$$Dose(i,j,k) = \sum_{(s,w)\in\mathcal{S}\times\mathcal{W}} t_{s,w} D_w(x_s, y_s, z_s, i, j, k), \tag{2.1}$$

where $D_w(x_s, y_s, z_s, i, j, k)$ is the dose delivered to the pixel $(i, j, k)$ by the shot of width $w$ centered at $(x_s, y_s, z_s)$.

To determine the form of $D_w$, the following procedure was followed. First, we simulated the delivery of a shot of width $w \in \mathcal{W}$, centered at the middle of the head of a previously scanned patient on the Gamma Knife. For each shot width, we determined the dose delivered in the $x$, $y$ and $z$ directions at given distances from the center of the shot from the simulation. The three values were then averaged to give a value of dose (for each width of shot) at a particular distance from the center, for example,

$$\bar{D}_w(d) = \frac{D_w(0,0,0,d,0,0) + D_w(0,0,0,0,d,0) + D_w(0,0,0,0,0,d)}{3}.$$

These values were used as data in a nonlinear parameter estimation problem.

The problem is reduced to determining a functional form for the dose delivered at a pixel that is a distance $d$ away from the center of the shot. A sum of error functions has been noted in the literature to approximate this dose distribution [**CKM98**]. We therefore used the following functional form

$$\sum_{i=1}^{2} \lambda_i \left(1 - \operatorname{erf}\left(\frac{d - r_i}{\sigma_i}\right)\right) \tag{2.2}$$

(with $d$ representing the distance) and fit the six parameters $\lambda_i$, $r_i$ and $\sigma_i$ to the data described above via least-squares, with different values for each width shot. In our work, we use the notation $\operatorname{erf}(x)$ to represent the integral of the standard normal distribution from $-\infty$ to $x$. The resulting nonlinear optimization problem

$$\min_{\lambda,r,\sigma} \left\| \bar{D}_w(d) - \sum_{i=1}^{2} \lambda_i \left(1 - \operatorname{erf}\left(\frac{d - r_i}{\sigma_i}\right)\right) \right\|^2$$

was solved using CONOPT. As can be seen by two representative pictorial representations of the resulting fits in Figure 6, the functional fit is very close to the observed data. The fit is best for the small shots (see Figure 6(a) for example), and decreases slightly in accuracy for the larger ones (see Figure 6(b) for example). The particular values of the parameters that we generated are given in Table 1. It is clear that a closer fit can be achieved using more parameters in (2.2). However, this will lead to more computational overhead in the optimization, which we believe is not necessary.

It is possible that the averaging in the $x$, $y$ and $z$ directions creates a small error for the particular patient at hand. Furthermore, there may be variations in dose across patients. These effects have been ignored in the subsequent formulations. In future work, along with investigations of the above issues, we may fit ellipsoids instead of spheres to the dose data since various researchers have commented that

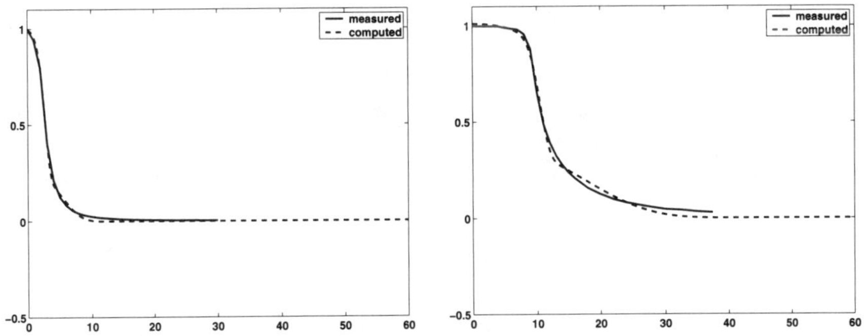


(a) 4mm shot

(b) 18mm shot

FIGURE 6. Nonlinear Parametric Fitting for Dose Data

| Shot | $\lambda_1$ | $r_1$ | $\sigma_1$ | $\lambda_2$ | $r_2$ | $\sigma_2$ |
|---|---|---|---|---|---|---|
| 4mm | 0.649200 | 1.365916 | 4.413680 | 0.599844 | 2.661771 | 0.668291 |
| 8mm | 0.401007 | 7.035785 | 5.702334 | 0.648584 | 4.849365 | 1.149176 |
| 14mm | 0.363704 | 13.97259 | 7.196694 | 0.657808 | 8.199979 | 1.321161 |
| 18mm | 0.381801 | 17.67857 | 8.194611 | 0.634696 | 10.31583 | 1.441725 |

TABLE 1. Parameter for Dose Model

the dose is skewed in certain directions. While cubic b-spline approaches may outperform the fit that we achieve here, it is currently unclear how to use such approaches within the GAMS modeling format that we employ.

## 3. Optimization Formulation

Once a description of the dose is determined, an optimization model can be formulated. The basic variables of the optimization include the number of shots of radiation that will be delivered, along with the width of the shot $w$, the coordinates of the center location of the shot $(x_s, y_s, z_s)$, and the time $t_{s,w}$ that each shot is exposed. In practice, we consider a grid of pixels, noting two subsets of this grid, namely $\mathcal{T}$ and $\mathcal{N}$, that represent the subset of pixels that are in (out) of the target respectively.

Neurosurgeons commonly use isodose curves as a means of judging the quality of a treatment plan. The 50% isodose curve is a curve that encompasses all of the pixels that receive at least 50% of that maximum dose that is delivered to any pixel in the patient. The neurosurgeon may wish to impose a requirement that the entire target is surrounded by an isodose line of x%. For example, a constraint that the 50% line must surround the target can be modeled by imposing strict lower and upper bounds on the dose allowed in the target, namely for all $(i, j, k) \in \mathcal{T}$

$$1 \leq Dose(i, j, k) \leq 2.$$

In this way, the 50% isodose curve is guaranteed to cover the target. The isodose requirements can be changed by simply modifying the numerical values of the bounds.

We have tested a variety of optimization formulations. For this paper, however, we will simply discuss two. The first formulation imposes a constraint on the minimum isodose line that must surround the target. Given this constraint, the goal is to minimize the dose outside of the target. The formulation is:

$$
\begin{array}{lrl}
\min & & \displaystyle\sum_{(i,j,k)\in\mathcal{N}} Dose(i,j,k) \\
\text{subject to} & Dose(i,j,k) = & \displaystyle\sum_{(s,w)\in\mathcal{S}\times\mathcal{W}} t_{s,w} D_w(x_s,y_s,z_s,i,j,k) \\
 & & 1 \le Dose(i,j,k) \le 2,\ \forall (i,j,k)\in\mathcal{T} \\
 & & t_{s,w} \ge 0 \\
 & & \operatorname{card}\left(\{(s,w)\in\mathcal{S}\times\mathcal{W} \mid t_{s,w}>0\}\right) \le n.
\end{array}
\tag{3.1}
$$

The final constraint states that no more than $n$ shots are to be used in the plan.

The second formulation uses a constraint to control the conformity of the plan. The constraint specifies that at least y% of the total dose must be deposited in the target. An upper bound is also placed on the dose to the target. Given these constraints the optimizer seeks to minimize the total underdosage in the target. A pixel is considered to be underdosed if it receives less than the prescribed isodose, which for the example formulation is assumed to be 1. We actually use the optimization process to model UnderDose. UnderDose is constrained to be no less that $\max(0, 1 - Dose)$ at every pixel in the target, and since we minimize UnderDose, it will take on the maximium of these two values at optimality. The complete formulation is:

$$
\begin{array}{lrl}
\min & & \displaystyle\sum_{(i,j,k)\in\mathcal{T}} UnderDose(i,j,k) \\
\text{subject to} & & Dose(i,j,k) = \sum_{(s,w)\in\mathcal{S}\times\mathcal{W}} t_{s,w} D_w(x_s,y_s,z_s,i,j,k) \\
 & & \dfrac{\sum_{(i,j,k)\in\mathcal{T}} Dose(i,j,k)}{\sum_{(i,j,k)} Dose(i,j,k)} \ge P \\
 & & 0 \le UnderDose(i,j,k) \ge 1 - Dose(i,j,k),\ \forall (i,j,k)\in\mathcal{T} \\
 & & Dose(i,j,k) \le 2,\ \forall (i,j,k)\in\mathcal{T} \\
 & & t_{s,w} \ge 0 \\
 & & \operatorname{card}\left(\{(s,w)\in\mathcal{S}\times\mathcal{W} \mid t_{s,w}>0\}\right) \le n.
\end{array}
\tag{3.2}
$$

In this formulation, $P$ is the fraction of the total dose that must be deposited in the target; typically we choose values for $P$ between 25% and 40%. For solution purposes, we rearrange the equation involving P so that it is a purely linear equation (i.e. multiply both sides by the term in the denominator). In practical application, rather than calculate the dose at every pixel, it is easy to accurately estimate the total dose delivered by a plan based solely on the $t_{s,w}$ variables and other pre-calculated constants.

Both of the formulations are based on the assumption that the neurosurgeon can use the volume of the target and the irregularity of its shape in order to a-priori determine a realistic upper bound $n$ on the number of shots required for treatment. Therefore, the model bounds the number of shots a priori, rather than forcing the model to minimize this number. Since most of the comments we make in the sequel apply equally well to either formulation, we will restrict our exposition mainly to (3.1) in the sequel solely for clarity.

Practical issues must also be considered. For example, a large number of shots can lead to a treatment time that is unacceptably long. This is because after each shot is delivered, the staff must enter the room and perform a series of adjustments. Also, if a change in the shot width is required, the user must exchange focusing helmets.

Several issues need to be resolved to create models that are practical, implementable, and solvable (in a reasonable time frame). Two main approaches are proposed in this paper, namely using mixed integer programming (MIP) and nonlinear programming (NLP).

**3.1. MIP approach.** Since the optimization technology associated with solving nonlinear mixed integer programming problems is very limited, we restrict attention to linear models and thus need to ensure that the dose calculation constraint (2.1) is linear. To do this, we need to have $D_w(\cdots)$ as data, and thus need to determine all possible $(x_s, y_s, z_s)$ before optimization. We therefore a-priori choose a grid of possible shot locations $\mathcal{G}_\mathcal{S}$, and precalculate $D_{s,w}(i,j,k) := D_w(x_s, y_s, z_s, i, j, k)$ for each grid location $s \in \mathcal{G}_\mathcal{S}$, every pixel $(i,j,k)$ and each width $w \in \mathcal{W}$ and use the optimization algorithm to decide whether or not to use a shot at a particular location. A major drawback of this approach is the enormous amount of data that is required. We will discuss approaches to overcome this in the sequel. It is clear that the only shots that will be considered in this approach are shots that lie within the target. This is because our grid is generated a-priori, and we can easily determine whether or not a shot lies within the target.

However, the mixed integer approach allows us to introduce a new binary variable $\psi_{s,w}$ (heuristically, use the shot $(s,w)$ or not) and impose the constraint

$$t_{s,w} \leq \bar{T}\psi_{s,w}$$

to force $\psi_{s,w} = 1$ whenever $t_{s,w} > 0$. Here $\bar{T}$ is an (easy to estimate) upper bound on the length of time that a particular shot can be exposed. Once these variables are introduced, we then write the limiting shot constraint as

$$\sum_{(s,w)\in\mathcal{G}_\mathcal{S}\times\mathcal{W}} \psi_{s,w} \leq n.$$

For the problem outlined in (3.1), the resulting mixed integer optimization problem is:

$$\begin{array}{lll} \min & & \sum_{(i,j,k)\in\mathcal{R}} Dose(i,j,k) \\ \text{subject to} & Dose(i,j,k) = & \sum_{(s,w)\in\mathcal{G}_S\times\mathcal{W}} t_{s,w} D_{s,w}(i,j,k) \\ & & 1 \le Dose(i,j,k) \le 2,\ \forall (i,j,k) \in \mathcal{T} \\ & & 0 \le t_{s,w} \le \bar{T}\psi_{s,w},\ \psi_{s,w} \in \{0,1\} \\ & & \sum_{(s,w)\in\mathcal{G}_S\times\mathcal{W}} \psi_{s,w} \le n. \end{array}$$

Note that we have replaced $\mathcal{N}$ with $\mathcal{R}$ in the above problem. In order to reduce the amount of data required and since the dose drops off rapidly away from its center, we only minimize the dose in a rind $\mathcal{R}$ around the target, instead of everywhere.

The above formulation is a large-scale mixed integer problem [**NW88, Wol98**] that is typically computationally intractable in the time allowed. Several methods are outlined below to further reduce the large amounts of data arising from the realistic dose calculations.

The benefits of reducing the possible shot locations $\mathcal{G}_S$ are two-fold. Firstly, the number of integer variables $\psi_{s,w}$ (typically the limiting computational factor for solution) is reduced. Secondly, the amount of precalculated data required is reduced. The following heuristic was used to attempt to reduce the number of shots as much as possible. We first generate a coarse grid of large shots and for a particular grid spacing and grid offset, we calculate the number of these shots that hit the target. By searching over all grid offsets, we determine the offset that maximizes the number of large shots from this grid that are on target. Having established this, we randomly place small shots near the boundary of the target until a prespecified number of shots is generated. It is an unresolved research issue to determine the proportion of large to small shots, and the number of possible locations that we should use.

There are several interesting features of the MIP approach. As noted above, there are large amounts of data, and many integer variables unless $\mathcal{G}_S$ is severely restricted. By restricting the grid too much, the resulting constrained problem is infeasible. Furthermore, even with the restrictions and a very fast MIP solver such as CPLEX, the problems take enormous amounts of time to solve or determine integer infeasibility. The major benefit of this approach, however, is that it provides a guarantee of global optimality for that particular choice of $\mathcal{G}_S$ and $\mathcal{R}$.

**3.2. Nonlinear Approach.** A very different approach is to use nonlinear programming and approximate the counting of shots. The basic idea is to use a migrating shot technique. This approach takes a limited number of shots, and the position each shot can change over the course of the optimization. Given we are using $n$ shots, we introduce variables $x_s$, $y_s$ and $z_s$ and use the nonlinear function formulation (2.2) for the dose calculation. Note that with this approach there are just $3n$ variables to specify shot location, and $n$ is typically small in the application (namely 5 - 20).

This has the key advantage over the MIP formulation that the "data" is much smaller. However, the nonlinear programming problems are still quite difficult to solve (they are not convex, and hence may have local solutions that are not global solutions). Secondly, while it is easy to limit the number of shot locations to $n$, the actual constraint requires that there are no more than $n$ shots are to be used in the plan. The missing feature is to choose a particular width of shot at each location.

To implement this feature, we use a two phase approach. Note that a shot of width $w$ is used at location $s$ if $t_{s,w} > 0$. The main idea is to approximate the step function $H(t) = 1$ if $t > 0$ and $H(t) = 0$ when $t = 0$ by a nonlinear function,

$$H(t) \approx H_\alpha(t) := \frac{2 \arctan \alpha t}{\pi}.$$

For increasing values of $\alpha$, $H_\alpha$ becomes a closer approximation to the step function $H$. Therfore, for the approach outlined with (3.1), we solve the following problem

$$\begin{array}{lll} \min & & \displaystyle\sum_{(i,j,k)\in\mathcal{R}} Dose(i,j,k) \\ \text{subject to} & Dose(i,j,k) = & \displaystyle\sum_{(s,w)\in\{1,\dots,n\}\times\mathcal{W}} t_{s,w} D_w(x_s, y_s, z_s, i, j, k) \\ & & 1 \le Dose(i,j,k) \le 2, \ \forall (i,j,k) \in \mathcal{T} \\ & & t_{s,w} \ge 0 \\ & & \sum_{w\in\mathcal{W}} H_\alpha(t_{s,w}) \le 1, \ \forall s \in \{1,\dots,n\}, \end{array}$$

with this approximation and $\alpha = 10$, then we fix any shots that have zero width to have $t_{s,w} = 0$, and reoptimize the following problem (with $\alpha = 100$)

$$\begin{array}{lll} \min & & \displaystyle\sum_{(i,j,k)\in\mathcal{R}} Dose(i,j,k) \\ \text{subject to} & Dose(i,j,k) = & \displaystyle\sum_{(s,w)\in\{1,\dots,n\}\times\mathcal{W}} t_{s,w} D_w(x_s, y_s, z_s, i, j, k) \\ & & 1 \le Dose(i,j,k) \le 2, \ \forall (i,j,k) \in \mathcal{T} \\ & & t_{s,w} \ge 0 \\ & & \sum_{(s,w)\in\{1,\dots,n\}\times\mathcal{W}} H_\alpha(t_{s,w}) \le n \\ & & t_{s,w} = 0, \ \forall (s,w) \in F. \end{array}$$

Here $F$ represents the subset of $\{1,\dots,n\} \times \mathcal{W}$ of variables that were fixed at 0 based on the solution of the first optimization. Note that both problems are highly nonconvex, so there is no guarantee of global optimality. We experimented with starting points somewhat and use equal values of $t$ for the results of the sequel.

Another disadvantage of the nonlinear approach is that it is difficult to constrain the location of the shots to within the target. In fact, we do not add such a constraint, but choose appropriate starting values for $(x_s, y_s, z_s)$ within the (interior of the) target to encourage the final shot locations to also be within the target. In all our experiments, this was indeed the case, although several shots can be located close to the boundary.

## 4. Computational Results

**4.1. Two-dimensional testing environment.** We created a testing environment for our models within MATLAB, using the image processing toolbox. The testing environment provided a means to generate two-dimensional "tumors" of various shapes and sizes and visualize the results of the optimization applied to these targets. The optimization models were written in the GAMS modeling language [**BKM88**], a high-level modeling system for mathematical programming problems. It consists of a language compiler and a variety of integrated high-performance solvers. We used the MATLAB/GAMS interface [**Fer98**] to communicate between these two software packages. GAMS was chosen for the optimizations based on its ability to formulate various types of models and then apply state-of-the-art optimization algorithms for their solution. In particular, for the mixed integer problems we used the CPLEX optimization package [**ILO**], whereas for nonlinear programs, CONOPT [**Dru85**] was used. While the two dimensional testing environment significantly reduces the amount of data present in the model, we found this to be an excellent testing tool to pinpoint the advantages and disadvantages of various formulations. In particular, it is very easy to change between a wide variety of model types, objective functions and constraints.

We present results for two cases. In the first instance, we prescribe 3 shots of radiation. The rind around the target has width 10. The resulting size of the target and rind is respectively 235 and 1092 pixels. The results of applying the mixed integer and nonlinear programming approaches are shown in Figure 7. The tumor outline is shown is cyan, while the scale of the intensities is given on the bar to the right of each figure. The MIP solution, shown in Figure 7(a), results in two 8mm and one 14mm shots, with a total dose to the rind of 510. The optimization took 165 seconds, terminating with a relative optimality tolerance of 1%. The size of the grid of potential shots was 150.

The NLP solution, shown in Figure 7(a), results in one 8mm and two 14mm shots, that combine to give a total dose to the rind of 490. The optimization took 65 seconds, somewhat faster than the MIP algorithm. Although the solution is not guaranteed optimal, it results in a comparable solution to the MIP code. Since the NLP code can choose the locations of the shots (as opposed to selecting one of the shots from the grid) it is possible to find better solutions using the NLP approach than the MIP approach.

To show the effects of more data on our approaches, Figure 8 exhibits the corresponding solutions for a larger case. Again, the target is outlined in cyan, this time containing 1008 pixels. The rind again has radius 10, resulting in 1365 pixels. We attempted to find solutions using seven shots.

The NLP solution, shown as Figure 8(b) took 1354 seconds to generate. It uses two 8mm, three 14mm and two 18mm, with a total dose to the enclosing rind of 1017. Note that the optimization time is slightly over the 20 minute limit, but can be reduced (for example) by reducing the number of pixels in the rind.

The MIP solution that is shown as Figure 8(a) consists of one 8mm, two 14mm and four 18mm, with a total dose to the disk of 1123, again with 150 potential locations for shots. The optimization was terminated after two hours with relative optimality tolerance 17%, having explored 15,000 nodes of the branch and bound tree. We believe that this demonstrates that more work is needed in this area before the MIP approach becomes a viable alternative to the NLP approach.

**4.2. Real patient (three-dimensional) data.** After determining which models gave satisfactory results, we switched to three dimensional models, based on real patient data. Currently, the mixed integer programming solvers are unable to process the real patient data in an acceptable time, so we present only nonlinear programming results in this section.

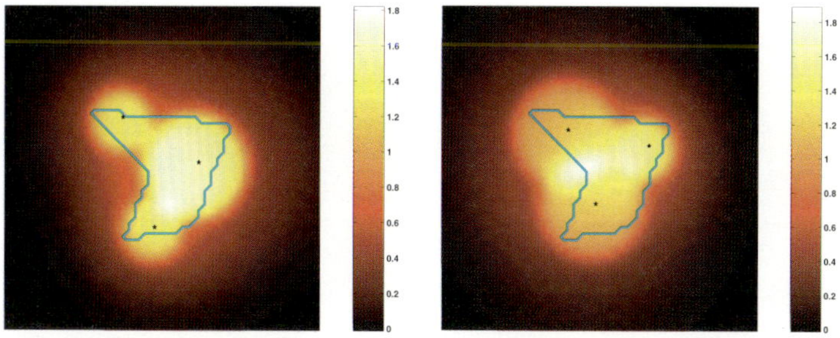

(a) MIP solution (b) NLP solution

FIGURE 7. Smaller Two-Dimensional Test Case

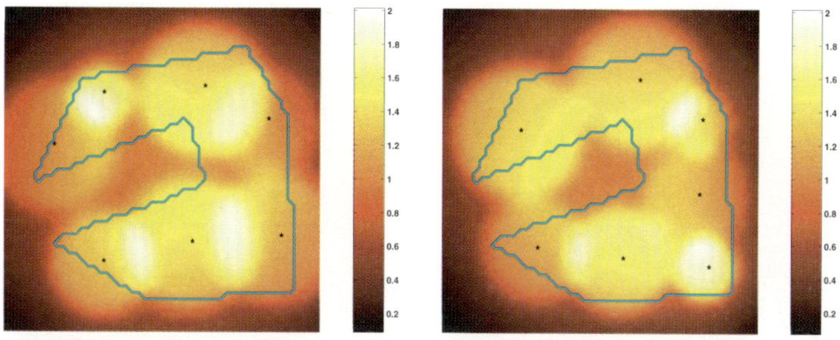

(a) MIP solution (b) NLP solution

FIGURE 8. Larger Two-Dimensional Test Case

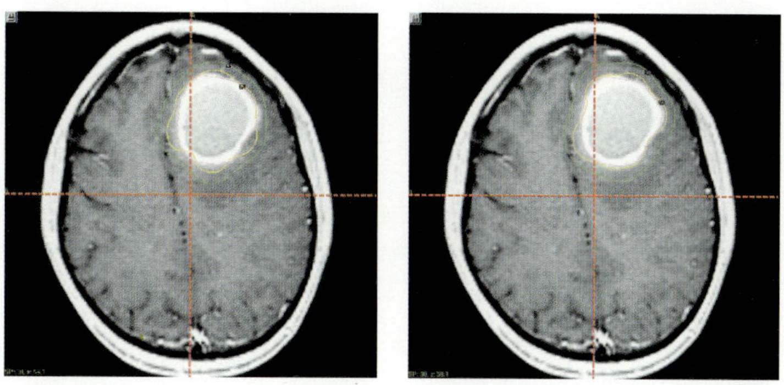

(a) Manual 13 shot solution (b) Optimized 13 shot solution

FIGURE 9. Isodose Curves for Patient 1; Axial Slice

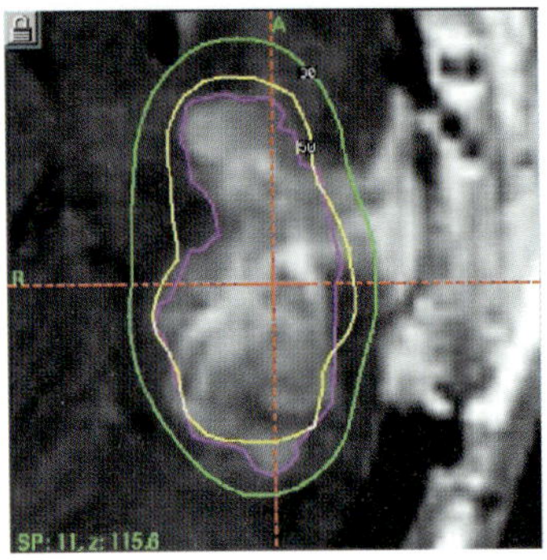


(a) Manual 7 shot solution;
Axial Slice

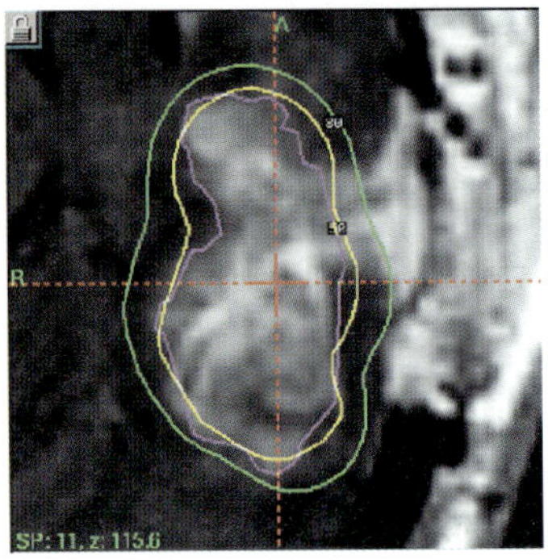


(b) Optimized 6 shot solution;
Axial Slice

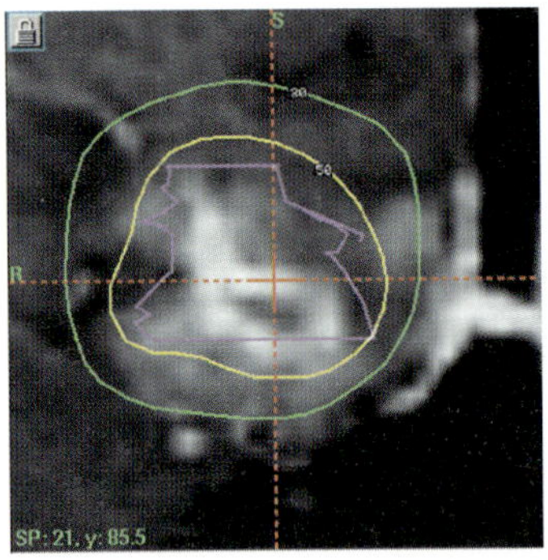


(c) Manual 7 shot solution;
Coronal Slice

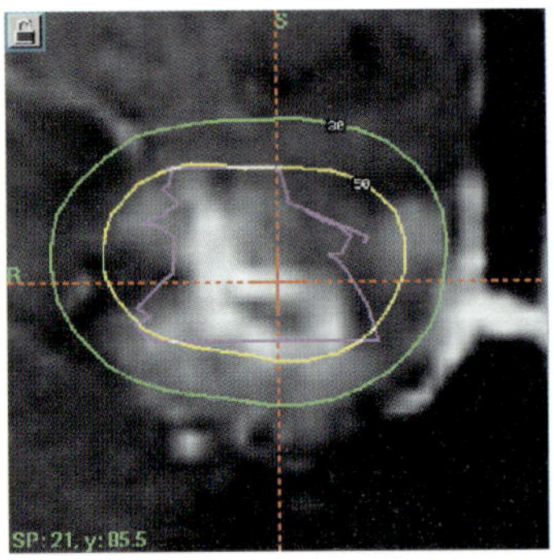


(d) Optimized 6 shot solution;
Coronal Slice

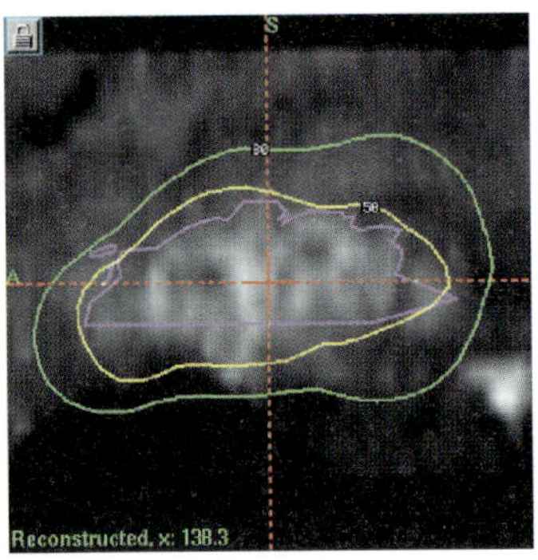


(e) Manual 7 shot solution;
Sagittal Slice

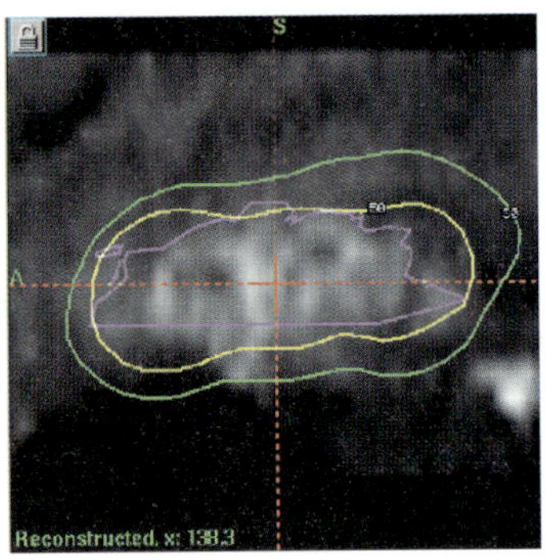


(f) Optimized 6 shot solution;
Sagittal Slice

FIGURE 10. Isodose Curves for Patient 2

The nonlinear programming approach to treatment planning has currently been applied to eight patient cases, and each optimized plan was compared to the original treatment plan produced by the neurosurgeon. These comparisons were based upon analysis of the dose-volume histogram, the minimum target dose and the the conformity index (the volume of the prescription isodose divided by the target volume). Treatment plans using 5-8 shots were typically optimized in less than 15 minutes on a Sparc Ultra-10 with a 330 MHz processor.

A treatment plan comparison is shown in Figure 9. Each image includes a pink line that indicates the border of the tumor. The 50% isodose line is shown in yellow, and the 30% isodose line is shown in green. Due to the massive size of this patient's tumor (35.3 cubic centimeters), a large number of shots were required in order to achieve adequate tumor coverage. In fact, this tumor is at the very limit of size of tumors that are treated using the Gamma Knife at the University of Maryland. Consequently, the treatment planning process was very tedious and required a great deal of artful manipulation on the part of the neurosurgeon.

The manual treatment plan (Figure 9a) used 13 shots of radiation, and is an excellent plan for such a large tumor. An optimized plan (Figure 9b) was produced using the same number of shots. A constraint guaranteed that the entire tumor volume was surrounded by the 47% isodose line, and the optimizer sought to minimize the total dose delivered outside of the tumor. The optimized plan is very similar in quality to the plan produced by an experienced nuerosugeon. In fact, the optimized plan increased the minimum target dose by 2% and provided a small improvement in the conformity index (1.28 versus 1.27). The purpose of including these results here is to show that an automatic plan can reproduce (or slightly improve) a plan determined by an expert in the field, for even the most difficult cases.

The results for a second patient are shown in Figure 10. In each figure, the purple line denotes the border of the tumor. The 50% isodose line is in yellow, and the 30% is line is in green. In this case, the manual treatement plan used 7 shots of radiation. The optimization was set up to use 6 shots of radiation. A constraint was included that specified that at least 30% of the total dose delivered to the patient had to be deposited in the target. Given this contraint, the optimizer sought to minimize the underdosage to the tumor (as in (3.2)). A voxel was considered to be underdosed if it received less than 50% of the maximum tumor dose. Again, the figures show the plan to be similar in quality, although in this case, the minimum tumor dose was increased from 43% to 44% and the conformity index was simultaneously improved from 1.70 to 1.75. Furthermore, the optimization process used one less shot of radiation than the expert.

The results shown in Figures 9-10 illustrate that this inverse treatment planning can be used to produce high-quality conformal treatment plans. A more detailed analysis of patient data will be provided in a future publication.

The robustness of the technique is of critical importance. For the eight patient cases optimized thus far, each shot was initially assigned a random location inside of the target. The initial guess incorporated uniform beam weights. Despite this somewhat arbitrary starting point, high-quality dose distributions have been obtained in all cases.

## 5. Conclusions and Future Work

Currently neurosurgeons produce suboptimal treatment plans by hand. We believe an automated approach based on optimization will generate not only more uniform, better treatment plans, but also produce them in less time than is currently used. The results given in this paper seem to indicate that the approach outlined here is sufficiently robust and quick to be used in a practical setting.

We believe it is important to integrate our procedure into the real system. Currently, we generate the target volume on the Gamma Knife and export it to the optimization program. The optimization program then generates a treatment plan that is imported back onto the Gamma Knife. This is somewhat time consuming, and likely to produce some data errors due to finite precision arithmetic and fixed size output. A full integration into the Gamma Knife planning system would reduce this time, and remove any of these errors.

We believe it is important to reduce the optimization times still further. This would give the ability to try "what if scenarios" at treatment planning time. For example, can conformality be improved if more shots are allowed? Can we treat the tumor at the 60% isodose level?

Apart from the time considerations, the outstanding question from an optimization viewpoint is the issue of global versus local optimality. While the nonlinear programming approach has proven very effective in practice, there is no guarantee that there is not a better treatment plan. Understanding how to improve the robustness of the model and solution techniques will enable more definite bounds to be given on the solution quality compared to the best possible solution.

## 6. Acknowledgements

We would like to thank Lijun Ma, Ph.D. and Cedric Yu, D.Sc., for their assistance with this project.

## References

[BKM88] A. Brooke, D. Kendrick, and A. Meeraus, *GAMS: A user's guide*, The Scientific Press, South San Francisco, CA, 1988.

[CKM98] P. S. Cho, H. G. Kuterdem, and R. J. Marks, *A spherical dose model for radiosurgery treatment planning*, Physics in Medicine and Biology **43** (1998), 3145–3148.

[Dru85] A. Drud, *CONOPT: A GRG code for large sparse dynamic nonlinear optimization problems*, Mathematical Programming **31** (1985), 153–191.

[Fer98] M. C. Ferris, *MATLAB and GAMS: Interfacing optimization and visualization software*, Mathematical Programming Technical Report 98-19, Computer Sciences Department, University of Wisconsin, Madison, Wisconsin, 1998.

[Gan97] J. C. Ganz, *Gamma knife surgery*, Springer-Verlag Wien, Austria, 1997.

[ILO] ILOG CPLEX Division, 889 Alder Avenue, Incline Village, Nevada, *CPLEX optimizer*, http://www.cplex.com/.

[LSY+99] L. Luo, H. Shu, W. Yu, Y. Yan, X. Bao, and Y. Fu, *Optimizing computerized treatment planning for the gamma knife by source culling*, International Journal of Radiation Oncology, Biology and Physics **45** (1999), no. 5, 1339–1346.

[NW88] G. L. Nemhauser and L. A. Wolsey, *Integer and combinatorial optimization*, John Wiley & Sons, New York, NY, 1988.

[SSV93] R. A. Stone, V. Smith, and L. Verhey, *Inverse planning for the Gamma Knife*, Medical Physics **20** (1993), 865.

[WB99] Q. J. Wu and J. D. Bourland, *Morphology-guided radiosurgery treatment planning and optimization for multiple isocenters*, Medical Physics **26** (1999), no. 10, 2151–2160.

[Wol98] L. A. Wolsey, *Integer programming*, John Wiley & Sons, New York, NY, 1998.

[YSB97] Y. Yan, H. Shu, and X. Bao, *Clinical treatment planning optimization by Powell's method for Gamma unit treatmen system*, International Journal of Radiation Oncology, Biology and Physics **39** (1997), 247–254.

COMPUTER SCIENCES DEPARTMENT, UNIVERSITY OF WISCONSIN, 1210 WEST DAYTON STREET, MADISON, WISCONSIN 53706

*E-mail address*: ferris@cs.wisc.edu

DEPARTMENT OF RADIATION ONCOLOGY, UNIVERSITY OF MARYLAND SCHOOL OF MEDICINE, 22 SOUTH GREEN STREET, BALTIMORE, MD 21201

*E-mail address*: dshep001@umaryland.edu

DIMACS Series in Discrete Mathematics
and Theoretical Computer Science
Volume **55**, 2000

# Sphere Packing Using Morphological Analysis

Q. Jackie Wu

**Abstract.** We present a solution to the three-dimensional (3D) sphere packing problem involving different sphere sizes by using morphological analysis of the target volume and shape. The medial axis transformation provides the optimal location and size of each sphere. The total number of spheres packed by this method is also minimal. The clinical application of such packing in Gamma Knife radiosurgery is also presented.

## 1. The Gamma Knife

The idea of stereotactic radiosurgery was introduced by Dr. Lars Leksell of the Karolinska Hospital in Stockholm, Sweden in 1951 (1). Stereotactic radiosurgery may be the only feasible treatment for some patients with surgically unresectable brain lesions, for those with residual lesions after attempted resection or for patients whose general health or advanced age precludes conventional intracranial surgery. Stereotactic radiosurgery delivers a single high dose of ionizing radiation to a radiographically well-defined, small intracranial target without delivering any significant fraction of the prescribed dose to the surrounding brain tissues. Sparing normal brain tissue is possible partially because of the steep dose gradient at the target periphery and partially because the minimum target isodose contour - often as low as 50% of the maximum dose - can be closely conformed to a 3-dimensional target configuration through the use of multiple isocenters (shots) of varying diameters. The main requirements for radiosurgery are: 1) Accurate determination of the target volume, usually with a spatial resolution of at least 1 mm; 2) Accurate three-dimensional dose

---

2000 *Mathematics Subject Classification, Primary 92C50, 58W99, 68R05.*
The author was in part supported by American Cancer Society under grant ACS – IRG #91-022.

calculation in real time; 3) Accurate spatial (within 1mm) delivery of the dose to the target volume; and 4) Homogeneous dose distribution inside the target volume and very sharp dose fall-off in regions surrounding the target volume.

One of the commonly used radiosurgery technique is the Gamma Knife. The Gamma Knife unit delivers a single, high dose of ionizing radiation emanating from 201 cobalt-60 unit sources (2-3). Figure 1 shows the diagram of a gamma knife unit. The central beam of the 201 source array lays at a fixed angle of $55^{\circ}$ to the horizontal plane. Each cobalt source has nominal 1TBq (30 Ci) activity with the diameter of the source at 1 mm and with a source-to-focus distance of approximately 40 cm. The sources lay in an arc +/- $48^{\circ}$ from the central beam along the long axis of the treatment table and +/- $80^{\circ}$ along the transverse axis to the table. In this design, the gamma-ray beams can enter the skull from almost 25% of all possible directions. The large solid angle covered by these beams has the advantage that the dose between the surface and the target can be distributed over a large volume of normal brain tissue.

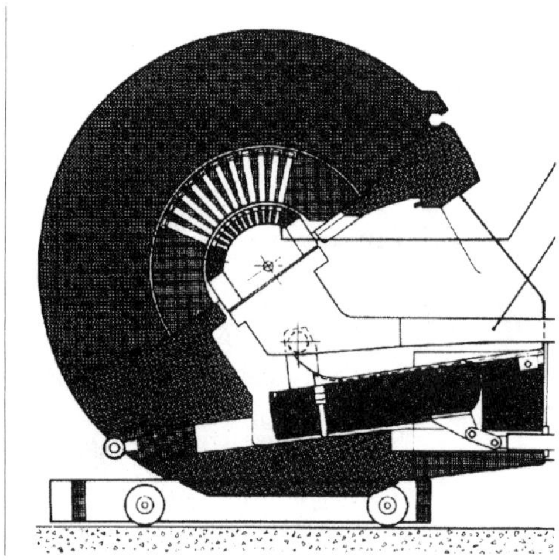

Figure 1. The Gamma Knife unit in treatment position.

All 201 beams simultaneously intersect at the isocenter, resulting in an approximately spherical dose distribution at the high dose levels. Irradiation of the isocenter to deliver a dose is termed a "shot". Shots can be represented as different spheres (3D) whose diameters are given as their 50% isodose levels (ISL). Four interchangeable outer collimator helmets, with beam channel diameters of 4, 8, 14 or 18 mm, are available for irradiating different sized volumes. For a target volume larger than one shot or for an irregular shape, multiple shots can be used to cover the entire target. In practice, most target volumes are treated with one to fifteen shots.

Because there are no moving parts during treatment, the precision of the dose delivery to the target resulting from the mechanical alignment, can be extremely high (within a fraction of 1 mm). The ultimate accuracy and precision of the dose delivery to the target volume is thus essentially limited by the precision of the target localization procedures and conformity of dose distribution to the defined target volume.

## 2. Treatment Planning of the Gamma Knife Radiosurgery

The current planning process is a manual trial-and-error approach that often requires several iterations. With very limited choices of shot size, multiple isocenters must be used to conform the global radiation dose distribution to match an often irregular tumor volume. "Packing" these shots together to generate a globally optimal dose distribution is a special combinatorial optimization problem. Denote the optimization process as $P_N(\bigcup_{j=1,\ldots,N} S^{d_j}(x_j))$, where N is the total number of shots used in the plan and $S^{d_j}(x_j)$ is shot j with its isocenter vector $x_j$ and size $d_j$. Note that the actual number of parameters in this optimization process is not defined. The location of each shot, its collimator size, and its relative weight as well as the total number of shots (N) being used are coupled to generate a conformal dose distribution. Hence, these parameters should be optimized simultaneously. Clearly, the irregularity of a target volume greatly influences the number and sizes of the shots used to conform to the target. A more irregular shape may demand a greater number of shots with small shot sizes, while a very rounded or ellipsoid target shape may only need a few large shots. In other words, the planning process is to configure a shot set with an appropriate total number of shots as well as an optimal position and size for each shot, taking into consideration the target volume and shape. Therefore, such “packing” is a morphometric process rather than a simple volumetric “packing”. The scenario is similar to filling a randomly shaped bottle with spheres – the goal is that small spheres fill the sharp corners and large spheres fill more open regions so that the total number of spheres is minimized while the bottle is maximally filled.

For each shot, there is an associated diameter which is defined as the range of its 50% ISL. If the dose distribution of a shot is simplified as a sphere with each point inside the sphere having a dose value of 100% and points outside having a dose value of zero, the treatment planning process can be modeled as a geometric "packing" problem. In this instance, all the spheres are considered as “non-elastic”. For real shots, this concept indicates that shots can only overlap at their 50% or lower ISLs.

For a region R $\subset R^3$, and V(R) denoting the geometric volume of region $R$, the problem is to find a minimum set of spheres, $S^{d_1}(x_1)$, $S^{d_2}(x_2)$, ... $S^{d_N}(x_N)$, with diameter $d_i$, centered at $x_i$, that form the optimal coverage of R:

$$P_{N_{opt}}(R) = P_N(R) \Big| N = min(n), \forall n, \frac{V(R) - V(P_n(R))}{V(R)} \leq \varepsilon \qquad (1)$$

with $\quad P_N(R) = \bigcup_{i=1}^{N} S^{d_i}(x_i)$,

and $\quad S^{d_i}(x_i) \subset R, \ i = 1,2,\ldots,N$,

ε is the tolerance of the uncovered target region. This tolerance is both necessary as well as reasonable for clinical practice. Since the dose distribution of a shot can be considered as a point spread function, geometrical gap regions between shots are defined as uncovered regions, but they may receive a sufficient dose as the summation of the lower doses (<50%) from the multiple shots surrounding them.

## 3. The Optimal Shot and the Skeleton Concept

Assume that a shot covers part of the target region denoted by A, as shown in Figure 2. If the position of this shot is ideal, then the shot should touch the contour at as many points as possible. If the radius is also ideal, then it should be the maximum radius to fit in this curvature. Any shots with a radius greater than this maximum radius, would protrude from the contour somewhere in at least one direction since the contour curvature is tighter than the radius. Indeed, matching a shot with a target contour is observed during the manual treatment planning process when shot sizes and positions are purposefully selected to "match" a specific part of a target. This manual "matching" process has two components: 1) positioning the shot in the center of the region; and 2) using the largest shot size possible to provide high dose coverage to the target boundary. To arrange shots in such a fashion, the characteristics of a target shape are used as represented by a function called medial axis transformation (4).

"Skeleton", also called "medial axis transform" or "symmetric-axis transform", is the name given to an object descriptor used in shape analysis and pattern recognition (5-7). There are certain classes of objects for which "skeleton" appears to be quite appropriate and meaningful, such as some biological shapes. In two-dimensional cases, the "skeleton" is often described (4) as the internal quench points of a "grass fire" started at the border of a region (Figure 3(a)). Suppose that fires are lighted simultaneously along the entire boundary of a region. Each point on the boundary is then considered as a fire source that grows by consuming the grass uniformly in all directions, thereby producing a disk-shaped wave front. When two fire fronts intersect, the fire is extinguished at the point of contact. The extinction points delineate the skeleton. The skeleton can also be defined as the locus of the centers of all the maximal internal disks wholly contained within the region but not wholly contained within any other internal disks (Figure 3(b)). A maximal disk is always tangential to the boundary of the region at least two points.

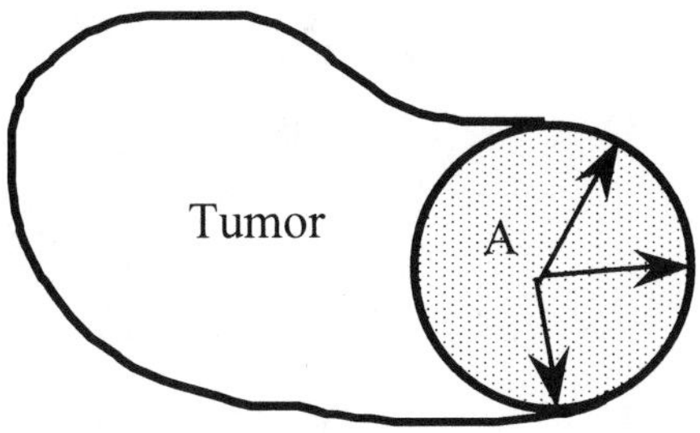


Figure 2. An ideally fitted shot.

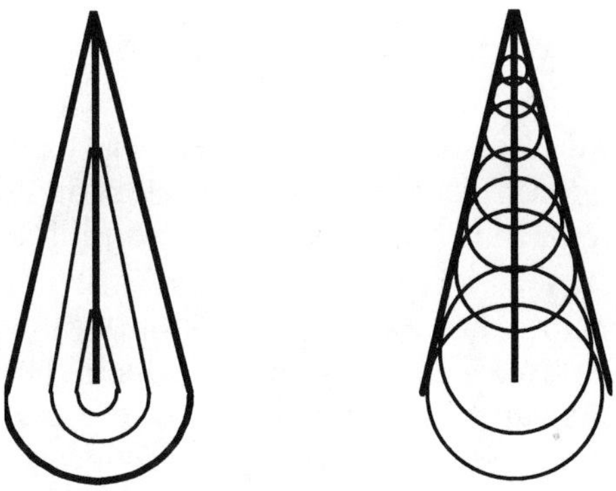


Figure 3. Skeleton definitions in 2D. (a) the grassfire representation (b) the internal maximum disk representation.

For a region $R$ in two-dimensional space, if the boundary is denoted as $\partial R$ and $X(R)$ is the corresponding skeleton, then $X(R)$ in two-dimensions can be defined as:

$$X(R)_{2D} = \{ x^* : x^* \in R \subset R^2, \exists x_1, x_2 \in \partial R,$$

$$\left\| x^* - x_1 \right\| = \left\| x^* - x_2 \right\| = \min_{x \in \partial R} \left\| x^* - x \right\| \Big\} \quad (2)$$

For the three-dimensional region $R$ in general, the corresponding skeleton $X(R)$ is defined as:

$$X(R)_{3D\text{-}line} = \Big\{ x^* : x^* \in R \subset R^3, \exists x_1, x_2, x_3 \in \partial R,$$

$$\left\| x^* - x_1 \right\| = \left\| x^* - x_2 \right\| = \left\| x^* - x_3 \right\| = \min_{x \in \partial R} \left\| x^* - x \right\| \Big\} \quad (3)$$

For certain regions, there is no sphere inside the region that can have more than two points touching the region boundary where the skeleton is degraded as a skeleton surface. In this case, the skeleton $X(R)$ is given as:

$$X(R)_{3D\text{-}surface} = \Big\{ x^* : x^* \in R \subset R^3, \exists x_1, x_2 \in \partial R,$$

$$\left\| x^* - x_1 \right\| = \left\| x^* - x_2 \right\| = \min_{x \in \partial R} \left\| x^* - x \right\| \Big\} \quad (4)$$

For each point $x^*$ on skeleton $X(R)$, there is an associated distance value, denoted as $l(x^*)$, which is the diameter of the maximal internal disk centered at $x^*$:

$$l(x^*) = Max(d) \big| S^d(x^*) \subset R. \quad (5)$$

In summary, the skeleton or the symmetric axis is a collection of intersecting curved line segments for a two-dimensional region $R$, and is a collection of intersecting curved lines and surfaces for a three-dimensional region $R$. A skeleton point with only one branch connected to it is called an end point, and a skeleton point with three or more branches connected to it is called a cross point.

Direct implementation of Equations (3) and (4) to search for the skeleton is generally prohibitive from a computational viewpoint because the search must calculate the distance from every interior point to every point on the boundary of the region. A new, fast 3D Euclidean distance skeletonization method is developed in this work and implemented as part of the optimization process (8).

## 4. The Optimization Method

Assume that the region R is a bounded region with a defined volume and surface. Each shot $S^{d_i}(x_i)$ centered at $x_i$ with diameter $d_i$ will cover a region $V(S^{d_i}(x_i))$. Since region R is bounded, there is a finite number of spheres that can cover the region within the coverage tolerance, $\varepsilon$.

Mathematically, there is a finite number (N) of shots $S^{d_i}(x_i)$, where i = 1, 2, … N, that can cover region R:

$$\frac{V(R)-V\left[\bigcup_{i=1}^{N} S^{d_i}(x_i)\right]}{V(R)} \leq \varepsilon \tag{6}$$

Consequently, for coverage V(R) in Equation (6), we can say that the center of each sphere is on the corresponding skeleton, i.e.,

$$\frac{V(R)-V\left[\bigcup_{i=1, x_i \in X(R_{i-1})}^{N} S^{d_i}(x_i)\right]}{V(R)} \leq \varepsilon, \tag{7}$$

where $R_i = R - \bigcup_{j=1}^{i} S^{d_j}(x_j)$

This can be proven by the skeleton definition. By extending Equation (7), an optimal coverage of target region $R$, $P_{\mathrm{Nopt}}(R)$, can be formulated as,

$$\begin{aligned} P_{N_{opt}}(R) &= \bigcup_{i}^{N_{opt}} S^{d_i}(x_i),\ x_i \in X(R), i = 1,2,\ldots N_{opt}, \\ &= P_N(R)\Big| N = min(n), \forall n, \frac{V(R)-V(P_n(R))}{V(R)} \leq \varepsilon, \end{aligned} \tag{8}$$

For shot coverage optimization of a target region, suppose a shot is made. The rest of the coverage is still an optimal coverage for the corresponding new region by excluding the region covered by the shot. This can be expressed as,

$$P_{N_{opt}}(R) = S^{d_i}(x_i) + P_{N_{opt}-1}(R - S^{d_i}(x_i)) \tag{9}$$

Actually, Equation (9) falls within the dynamic programming scheme. The key characteristics of dynamic programming are that no matter what previous states and strategies the process has been through, the optimal solution for the current stage is solely dependent on the current state. Therefore, no matter how many shots are made and no matter where they are located, if the current region to be optimized is the same, the process should reach the same optimal solution for the current region. More directly, if the first shot is an optimal shot and the region it covered is excluded, the process will still reach an optimal solution for the rest of the region, i.e. the optimal solution reached sequentially as above is the same as the optimal solution reached by simultaneously optimizing all the shots. Mathematically,

$$\min N \mid P_N(R)$$

$$= 1 + \min N \mid P_N(R - S^{d_i}(x_i)), \forall S^{d_i}(x_i) \subset P_{N_{opt}}(R) \tag{10}$$

Equation (10) states that the original optimization problem can be transferred to several sub problems by cutting spheres sequentially from the original region. For each sub problem, Equation (10) applies again and results in a recursive search process. It attempts all possible solutions and identifies the optimal one. Although Equation (10) does not explicitly require a sphere to be on the corresponding skeleton, the inherent characteristics of skeletonization dictates that any optimal sphere must be on the skeleton of that part of the region.

## 5. Results and Discussion

Figures 4 shows the resulting shot geometry or configuration and the corresponding dose distribution for an irregularly shaped target. The size of a shot, shown by a circle representing its 50% ISL, is a combined effect of collimator size and the relative weight assigned to it. Regions in gray in Figure 4(a) are the regions not covered by the shots, but these regions are not necessarily under-dosed. These regions are spread evenly throughout the target so that extremely "cold" spots are prevented. Figure 4(b) shows the dose distributions generated from gamma knife dose calculation software.

The dose from each shot is somewhat like a point spread function. Thus, there is always a transient region just outside the shot boundary (its 50% ISL) where the dose falls off radially. The dose distribution of these transient regions from all the shots overlaps slightly such that the dose in these regions will increase but will not exceed 100%. Thus, the final 50% ISL is a smoothed fusion of all individual 50% ISLs. Also of note is the cold spot inside the target in Figure 4(a); which is surrounded by three large shots and one small shot. Even though this area is relatively far from any of the four shots and the dose from each of the shots is less than 20%, the final dose is still within the requirement level.

Figure 4(a) shows a morphological dilation performed on the original target boundary so that the dilated target region is 2 mm larger around the boundary. The optimization proceeds on the dilated target region, and all the shots are confined to this region even though some of the shots may protrude outside the original target contour. The dilation is an effort to force the final 50% ISL to fully cover the target contour, as shown in Figure 4(b).

Skeletonization mimics manual treatment planning and optimization for radiosurgery, as the planner always attempts to center a shot in a region of coverage and uses the shot size to match the region. The optimization predicts the final dose distribution, and the result agrees well with the dose calculation. The majority of calculation time is consumed by skeletonization; therefore fast skeletonization algorithms will increase the optimization speed dramatically.

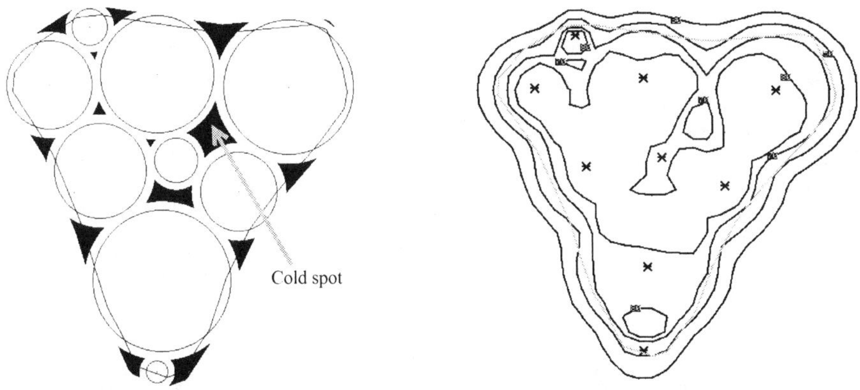


(a) (b)

Figure 4. Results. (a) the optimal plan, circles indicate shot locations and sizes, and (b) dose distribution of the plan; the gray line is the target contour, and the black lines are the 30%, 50%, 70%, and 80% ISLs.

## References


1. Leksell, L. 1951, "The stereotactic method and radiosurgery of the brain", *Acta Chir Scand*, 102:316-319
2. Wu, A. 1992, "Physics and dosimetry of the Gamma knife", *Neurosurg Clin North Am,* Vol 3(1):35-50
3. Walton. L., Bomford, C. K., Ramsden, D., 1987, "The Sheffield stereotactic radiosurgery unit: physical characteristics and principles of operation", *Br J Radiol,* 60:897-960
4. Blum, H. 1973, "Biological shape and visual science: Part I," *J Theor Biol*, 38, 205-287
5. Russ, J. C. 1989, "Uses of the Euclidean distance map for the measurement of features in images," *J Comput Assist Microsc,* 1(4), 343-375

6. Russ, J. C. 1990, Computer-Assisted Microscopy, Plenum Press.
7. Haralick, R. M. 1992, Computer and Robot Vision, Vol. 1, Chapter 5.
8. Wu, Q. J., Bourland, J. D., "3D skeletonization and its application in radiation therapy," *(In press) Comput Med Imag Graph.*

Department of Radiation Oncology, Case Western Reserve University
School of Medicine, Cleveland, OH 44106
*Email address: qjw@po.cwru.edu*

DIMACS Series in Discrete Mathematics
and Theoretical Computer Science
Volume 55, 2000

# Transition to Epileptic Seizures: Optimization

L. D. Iasemidis, D.-S. Shiau, J. C. Sackellares, and P. Pardalos

ABSTRACT. Epilepsy is a dynamical disorder of the brain. The existence of an epileptogenic focus that progressively entrains normal brain areas results to transitions of the brain from chaotic to less chaotic spatiotemporal states, the epileptic seizures. Measures of chaos of the electrical activity (EEG) of the brain (e.g. Lyapunov exponents) can be used to quantify this entrainment. Using optimization theory, in particular quadratic integer programming, we were able to select the brain sites that are entrained to each other and subsequently follow their entrainment over long time periods in one patient with 24 seizures. This procedure, which is applied to epilepsy research for the first time, resulted to identification of the epileptogenic focus and showed the possibility of prediction of epileptic seizures well in advance of their occurrence.

## 1. Introduction

Epilepsy is a common disorder of the brain, affecting approximately 1% of the population. It is characterized by intermittent seizures, that is, intermittent paroxysmal rhythmic electrical discharges within the cerebrum that disrupt normal brain function. Approximately 25% of patients with epilepsy have seizures that are resistant (refractory) to medical therapy. For some types of epilepsy (i.e. focal or partial epilepsy), there is a localized structural change in neuronal circuitry within the cerebrum which produces organized quasi-rhythmic discharges. These discharges then spread from the region of origin (epileptogenic zone) to activate other areas of the cerebral hemisphere. While the macroscopic and microscopic features of the epileptogenic zone have been described, the mechanism by which these fixed disturbances in local circuitry produce intermittent disturbances of brain function is not understood.

The most common type of epilepsy in adults is temporal lobe epilepsy. In this type of epilepsy, the temporal cortex, limbic structures and orbitofrontal cortex appear to play a critical role in the onset and spread of seizures. Temporal lobe seizures usually begin as paroxysmal electrical discharges in the hippocampus and often spread first to ipsilateral, then to contralateral cerebral cortex. These abnormal discharges result in a variety of intermittent clinical phenomena, including

2000 *Mathematics Subject Classification.* Primary 68T10, 62N01; Secondary 90C90.

This research is supported by a RO1 NS 31451 grant (the U.S.A. National Institute of Health), a V.A. Merit grant (the U.S.A. Department of Veterans Affairs), a Whitaker Foundation seed grant, and DBI 98-08210, EIA 98-75509 (U.S.A. National Science Foundation) grants.

motor, sensory, affective, cognitive, autonomic and psychic symptomatology. In medically refractory cases, surgical excision of the seizure focus may be effective means of seizure control. In those surgical candidates, continuous (24-hour) electroencephalographic recordings (EEG) are obtained for diagnostic purposes from subdural electrodes placed over the frontal and temporal cortex and depth electrodes implanted in the hippocampi bilaterally [**Nie87**]. Such recordings, typically including 4 to 10 clinical seizures per patient, offer a unique opportunity for research into the electrophysiological processes of epileptogenesis in man. Our group claims the biggest data base of such continuous multichannel EEG recordings in the world to date in terms of the number of epileptic patients and length of recording per patient.

Over the past several years, we have sought to: 1) find characteristic dynamical features in the EEG generated by the epileptogenic focus, 2) use the above features from multichannel EEG recordings over the human brain to identify the epileptogenic focus and 3) predict impending epileptic seizures. The possibility that such features might exist is suggested by the fact that there are structural abnormalities in the epileptogenic hippocampus (e.g., neuronal loss, dendritic abnormalities, and axonal sprouting of dentate granule cells) [**FSC64, MC66, MGHS91, BSHR95**] and localized zones of hypometabolism detectable with PET scans which are present during the interictal state, that is in the period between epileptic seizures [**EKPM82, ASSG87, SSAH90**]. Using techniques developed within the framework of nonlinear dynamics to identify nonlinearities in the EEG, signals generated by the epileptogenic focus and spike foci located in the temporal cortex contralateral to the epileptogenic focus have been found to be strongly nonlinear [**BD86, IZSW88, ISS93, IOSS94, SIZW89a, SIZW89b, LE95, CISR96, CIGR97, FMWB99, VVHP99, LMAV00**]. These findings suggested that the framework of nonlinear dynamics is suitable for investigation of the epileptic brain.

Traditionally, the initial occurrence of the characteristic focal rhythmic EEG discharge, most commonly in the hippocampus or mesial temporal cortex, is considered to be the onset of a seizure. However, through analysis of the spatiotemporal dynamics of such EEG recordings in patients with medically intractable temporal lobe epilepsy, we have discovered a preictal transition that precedes seizures for periods on the order of minutes to hours [**Ias91, IS91, IS96, IPS96, IPCG97**]. This preictal dynamical transition is characterized by a progressive convergence of measures of chaos (maximum Lyapunov exponents, i.e. $L_{max}$ values) at specific anatomical areas. The existence of a preictal transition period that can be detected by its quantitative dynamical characteristics, recently confirmed by other groups [**EL98, LE98, MALB98, LMBV99**], offers the possibility of predicting an impending seizure in time to intervene therapeutically in order to disrupt the preictal transition and prevent a seizure from occurring. For example, this disruption might be achieved by electrical or magnetic stimulation (e.g. vagal nerve stimulation) or by timely release of an anticonvulsant drug.

In this paper, the question of how to select specific brain areas from dynamical analysis of the EEG in order to maximize the detection of the preictal transition to epileptic seizures is addressed. In section 2, measures from nonlinear dynamics and results of their application to EEG in epilepsy are presented. In section 3, the methodology we developed for the selection of the brain sites is described. Results for detection of the preictal transition, and localization of the epileptogenic focus from the application of this method to EEG data, are presented in section 4. These results are discussed in the final section 5.

## 2. Background

### 2a. Nonlinear dynamical measures

Since its discovery by Richard Caton [**Cat75**] and its first systematic investigation by Hans Berger [**Ber29, Glo69**], the electroencephalogram has been the most utilized signal to clinically assess brain function. Unfortunately, traditional signal processing theory, based on very simple assumptions about the system that produces the signal (e.g. linearity assumption) has met the challenge of quantification of EEG with varying degrees of success. We can say that today the trained electroencephalographer or neurologist are still the golden standards in characterizing phasic events in the EEG such as spikes, the quantification of background activity (e.g. in sleep staging) and identification and localization of epileptic seizures. EEG characteristics such as alpha activity and seizures, instances of bursting behavior during light sleep, amplitude dependent frequency behavior (the smaller the amplitude the higher the EEG frequency) and existence of frequency harmonics (e.g. under photic driving conditions) are extensively reported in the clinical literature. However, all these characteristics belong to the long catalog of typical properties of nonlinear systems [**Jan91**].

The EEG, being the output of a multidimensional system, has statistical properties that depend on both time and space [**Lop87**]. Components of the brain (neurons) are densely interconnected and the EEG recorded from one site is inherently related to the activity at other sites. These components may functionally interact at different time instants. This makes the EEG a multivariable non-stationary time series. A well-established technique for visualizing the dynamical behavior of a multivariable system is to generate a phase space portrait of the system. A phase space portrait is created by treating each time-dependent variable of the system as a component of a vector in a multidimensional space, usually called state or phase space of the system. Each vector in the phase space represents an instantaneous state of the system. These time-dependent vectors are plotted sequentially in the phase space to represent the evolution of the state of the system over time. For many systems, this graphic display creates an object confined over time to a sub-region of the phase space. Such sub-regions of the phase space are called "attractors". The geometrical properties of these attractors provide information about the steady state of the system.

One of the problems in analyzing multidimensional systems is to decide which observable of the system (variable that can be measured) to analyze. In addition, experimental constraints may limit the number of observables that can be obtained. It turns out that when the variables of the system are related over time, which must be the case for any dynamical system to exist as a system, proper analysis of a single observable can provide information about the variables of the system that are related to this observable. In principle, through the method of delays described by Packard et al. [**PCFS80**] and Takens [**Tak81**], sampling of a single observable over time can approximate the position (state) of the system in a space spanned by the system variables related to this observable. This technique for the reconstruction of the phase space from one observable can be used for the EEG. In this case, a multidimensional phase space can be created from a single-electrode

EEG recording. Thus, in such an embedding, each state is represented in the phase space by a vector $x(t)$ whose components are the delayed versions of the original single-channel EEG time series $u(t)$, that is:

$$x(t) = [u(t), u(t-\tau), ..., u(t-(p-1)\times\tau)] \tag{2.1}$$

where $x(t)$ is the vector in the phase space at time $t$, $\tau$ is the time delay between successive components of $x(t)$, and $p$ is the embedding dimension of the reconstructed phase space.

The geometrical properties of the phase portrait of a system can be expressed quantitatively using measures that ultimately reflect the dynamics of the system. For example, the complexity of an attractor is reflected on its dimension [**GP83a, GP83b**]. The larger the dimension of an attractor, the more complicated it appears in the phase space. It is important to distinguish between the dimension of the phase space (embedding dimension) and the dimension of the attractor. The embedding dimension $p$ is the dimension of the phase space that contains the attractor and it is always a positive integer. On the other hand, the attractors' dimension $D$ may be a positive non-integer (fractal). $D$ is directly related to the number of variables of the system and is inversely related to the existing coupling among them. According to Takens, the embedding dimension $p$ should be at least equal to $(2D+1)$ in order to correctly embed an attractor in the phase space. Of the many different methods used to estimate $D$ of an object in the phase space, each has its own practical problems [**ERKS94, GSS91**]. The measure most often used to estimate $D$ is the phase space correlation dimension $\nu$. Methods for calculating $\nu$ from experimental data have been described [**May86, Kos92**] and were employed in our work to approximate $D$ of the epileptic attractor. In the EEG data we have analyzed to date, $\nu$ is found to be between 2 and 3 during an epileptic seizure. Therefore, in order to capture characteristics of the epileptic attractor, we have used an embedding dimension $p$ of 7 for the reconstruction of the phase space.

The entropy $H$ of an attractor embedded in $p$ dimensions also can be estimated from the phase space correlation integral [**GP83a, GP83b**] and is then related to the second order Renyi information, a lower bound to Shannon information [**Ren70**]. This entropy is mainly a $p-$dimensional joint entropy and it can be approximated in units of bits by $H_p(\epsilon) = -\log C_p(\epsilon)/\log 2$, where $\epsilon$ is the resolution of search in the phase space and $C_p$ is the phase space correlation integral. In this correspondence, for the estimation of $H$ we have used $p = 7$ and $\epsilon = \sigma/2$, where $\sigma$ is the standard deviation of the raw EEG data in each data segment.

An attractor is chaotic if, on the average, orbits originating from similar initial conditions (nearby points in the phase space) diverge exponentially fast (expansion process). If these orbits belong to an attractor of finite size, they will fold back into it as time evolves (folding process). The result of these two processes may be a stable, topologically layered, attractor [**Hol86**]. When the expansion process overcomes the folding process in some eigendirections of the attractor, the attractor is called chaotic. The measures that quantify the chaoticity of an attractor are the Kolmogorov entropy ($K$) and the Lyapunov exponents, typically measured in bits/sec [**Ose68, Pes77, Wal82**]. For an attractor to be chaotic, the Kolmogorov entropy or at least the maximum Lyapunov exponent ($L_{max}$) must be positive. The Kolmogorov (Sinai or metric) entropy ($K$), measures the uncertainty about the future state of the system given information about its previous states in the phase space. The Lyapunov exponents measure this average uncertainty along the local eigenvectors of an attractor in the phase space. If the phase space is of $p$ dimensions, we can estimate theoretically up to $p$ Lyapunov exponents. However, as expected, only $([D]+1)$ of them will be real. The rest will be spurious [**Aba96**]. Methods for calculating these dynamical measures from experimental data have been published [**Aba96, WSSV85, ISZW90, EKRC86**]. The estimation of the largest Lyapunov exponent in a chaotic system has been shown to be more reliable and reproducible than the estimation of the remaining exponents [**GSS91, VK86**], especially when $D$ is unknown and changes over time, as it is the case with high-dimensional and nonstationary data.

Since the brain is a nonstationary system, algorithms used to estimate measures of the brain dynamics should be capable of automatically identifying and appropriately weighing existing transients in the data. The method we developed for estimation of $L_{max}$ for nonstationary data, called $STL$ (Short Time Lyapunov), is taking in consideration possible nonstationarities in the EEG and is explained in detail in Iasemidis et al. [**IS91, ISZW90**]. We apply the $STL$ algorithm to EEG tracings from several brain sites, to create a set of $L_{max}$ time series containing local (in time and in space) information about the brain as a dynamical system [**IPS00**]. It will be shown that it is at this level of spatiotemporal analysis that reliable detection of the transition to epileptic seizures long prior to their actural occurrences is derived.

### 2b. Temporal dynamical analysis

Figure 1 shows our 28 electrode montage used for subdural and depth EEG recordings. We have typically analyzed continuous EEG signals for at least 1 hour before to 1 hour after a seizure, sampled with a sampling frequency of 200 $Hz$ and low-pass filtered at 70 $Hz$. In Figure 2 we depict typical EEG recordings from representative electrode sites of each area of recording, at the onset of an epileptic seizure, from a patient with temporal lobe seizures of right hippocampal origin (RTD). In Figure 3(a), a 10 sec ictal EEG signal at the seizure's onset from a single electrode (RTD2) in the epileptogenic zone is shown. The phase space portrait of this signal is shown in Figure 3(b). The correlation dimension ($\nu$) of the attractor shown is 2.5. The characteristics of the formed epileptic attractor in the phase space are typical for all of the seizures we have analyzed to date.

By dividing the recorded EEG data at an electrode site into non-overlapping segments, each 10.24 sec in duration, and estimating $H$ and $STL_{max}$ for each of these segments, profiles of $H$ and $STL_{max}$ over time are generated. A typical plot of $STL_{max}$ at an electrode site (RTD2) from the epiletogenic zone over time is shown in Figure 4(a). The exponent $STL_{max}$ is positive during the whole period of recording, 94 minutes preictally (prior to seizure $\sharp$ 10 onset), 2 minutes ictally (during the seizure) and 14 minutes postictally (after seizure's end). The seizure's onset corresponds to the maximum drop in the values of $STL_{max}$, thus the seizure can be detected from the lowest values of $STL_{max}$. However, ictal $STL_{max}$ is still positive implying a chaotic state even during the seizure. This is consistent with an interpretation of the seizure being a chaotic state with fewer degrees of freedom than before. It appears that these degrees of freedom are usually enough for the system to find a way out and the patient to recover. Comparing the mean value of $STL_{max}$ in the preictal state with the one in the postictal state, we can say that the preictal state is less chaotic than the immediate postictal one. In the preictal

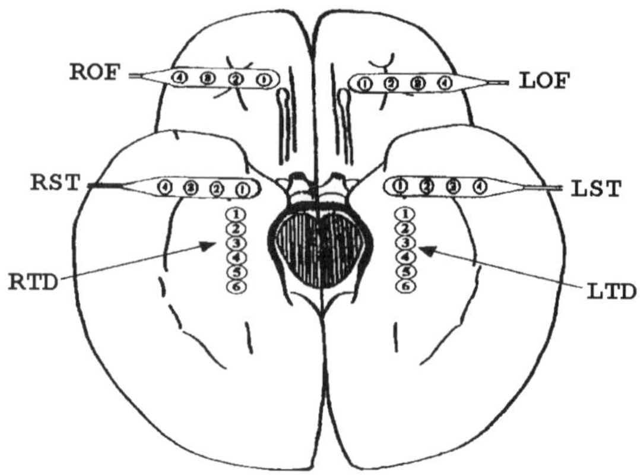


FIGURE 1. Schematic diagram of the depth and subdural electrode placement. This view from the inferior aspect of the brain shows the approximate location of depth electrodes, oriented along the anterior-posterior plane in the hippocampi (RTD - right temporal depth, LTD - left temporal depth), and subdural electrodes located beneath the orbitofrontal and subtemporal cortical surfaces (ROF - right orbitofrontal, LOF - left orbitofrontal, RST- right subtemporal, LST- left subtemporal).

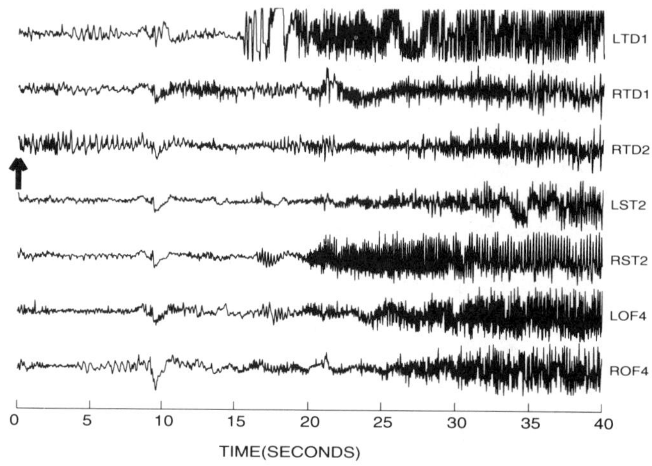


FIGURE 2. A 40 sec EEG segment at the onset of a right temporal lobe epileptic seizure. The ictal discharge begins in the right depth electrode (hippocampal RTD2) as a series of high frequency repetitive spikes (first 10 sec into the record), then spreads to RTD1 (next 10 sec into the record) and subsequently to the rest of the illustrated brain sites, namely left hippocampus (LTD1), right subtemporal (RST2), left orbitofrontal (LOF4), right orbitofrontal (ROF4) and finally left subtemporal (LST2) about 20 sec after the initial ictal onset at RTD2.

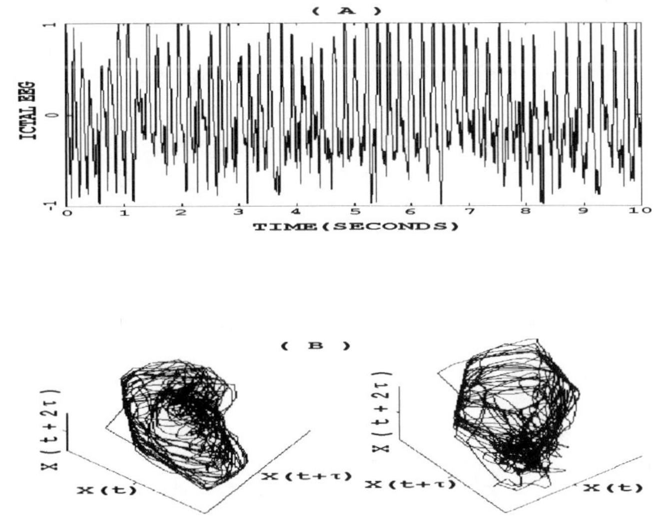


FIGURE 3. (a) A 10 sec segment of raw EEG data during an epileptic seizure recorded from electrode RTD2 in a patient with right temporal lobe epilepsy. (b) The epileptic attractor - EEG data in the reconstructed phase space with embedding dimension $p = 3$ and time delay $\tau = 20$ msec, in two different orientations.

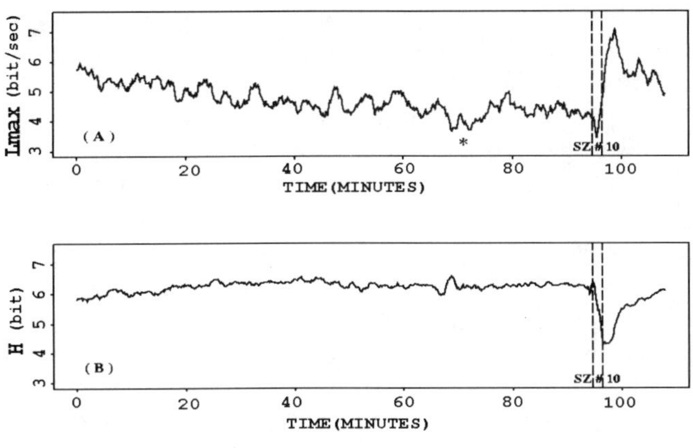


FIGURE 4. Smoothed $STL_{max}$ (a) and $H$ (b) profiles over 2 hours derived from an EEG signal recorded at RTD2. A seizure (SZ ♯ 10) startes and ended between the two vertical lines. The estimation of the $H$ and $L_{max}$ values was made by dividing the signal into non-overlapping segments of 10.24 sec each, using $p = 7$ and $\tau =$ 20 msec for the phase space reconstruction. The smoothing was performed by a 10 point (1.6 minutes) moving average window over the generated $STL_{max}$ and $H$ profiles.

state depicted in Figure 4(a), one can notice a trend of $STL_{max}$ toward lower values over the whole preictal period of seizure ♯ 10 with one prominent drop in the value of $STL_{max}$ at about 24 minutes prior to the seizure (denoted by an asterisk in the figure). This preictal drop in $STL_{max}$ can be explained as an attempt of the system toward a phase transition long before the actual seizure [**IS91**]. As we will show below, such an explanation becomes evident only after the consideration of the dynamics of the EEG at other electrode sites too. The entropy $H$ profile at RTD2 is shown in Figure 4(b). We note that, after seizure's onset, $H$ values drop later than $STL_{max}$ values.

### 2c. Spatiotemporal dynamical analysis

In this section, we quantify the spatiotemporal dynamical changes in the EEG. We employ the $T$-index (from the well-known t-test for comparisons of means of paired observations) as a measure of distance between the mean values of pairs of $STL_{max}$ profiles over time. The $T$-index at time $t$ between electrode sites $i$ and $j$ is defined as:

$$T_{i,j}(t) = \sqrt{N} \times E\{STL_{max,i} - STL_{max,j}\}/\sigma_{i,j}(t) \tag{2.2}$$

where $E\{\cdot\}$ is the sample average differences for the $(STL_{max,i} - STL_{max,j})$ estimated over a moving window $w_t(\lambda)$ defined as:

$$w_t(\lambda) = \begin{cases} 1 & \text{if } \lambda \in [t-N-1, t] \\ 0 & \text{if } \lambda \notin [t-N-1, t], \end{cases}$$

where N is the length of the moving window. Then, $\sigma_{i,j}(t)$ is the sample standard deviation of the $STL_{max}$ differences between electrode sites $i$ and $j$ within the moving window $w_t(\lambda)$. The thus defined $T$-index follows a $t$-distribution with N-1 degrees of freedom. For the estimation of the $T_{i,j}(t)$ indices in our data we used $N = 30$ ( i.e., average of 30 differences of $STL_{max}$ exponents between site $i$ and $j$ per moving window of approximately 5 minute duration). Therefore, a two-sided $t$-test with $N-1(=29)$ degrees of freedom, at a statictical significance level $\alpha$ should be used to test the null hypothesis, $H_o$: "brain sites $i$ and $j$ acquire identical $STL_{max}$ values at time t". If we set $\alpha = 0.05$, the probability of a type I error, or better, the probability of falsely rejecting $H_o$ if $H_o$ is true, is 5%. For the $T$-index to pass this test, the $T_{i,j}(t)$ value should be within the interval [-2.045,+2.045].

In Figure 5(a), preictal entrainment (long before the occurrence of a seizure) and postictal disentrainment (after the occurrence of the seizure) of the $STL_{max}$ profiles at two brain sites is shown. In Figure 5(b), this behavior is quantified by the $T$-index profile between these sites. From this figure, it is clear that attempts for a spatiotemporal entrainment between brain sites occur long before an epileptic seizure (first attempt about 73 minutes prior to seizure - see arrow). Postictally, this entrainment is reset. In Figure 6(a), the corresponding profiles for entropy $H$ between the same brain sites of Figure 5 are shown. $H$ profiles appear smoother than $STL_{max}$ profiles. It is noteworthy to mention that preictal entrainment is visible in the $H$ profile (first attempt about 30 minutes prior to seizure, see Figure 6(b)) much later than entrainment in the $STL_{max}$ profiles.

The challenge for optimal detection of the preictal transition is to appropriately select the brain sites that participate in the preictal entrainment (critical sites). In the following section we propose a method to select these critical sites based on global minimization of the average distance of entropy rates ($STL_{max}$) among

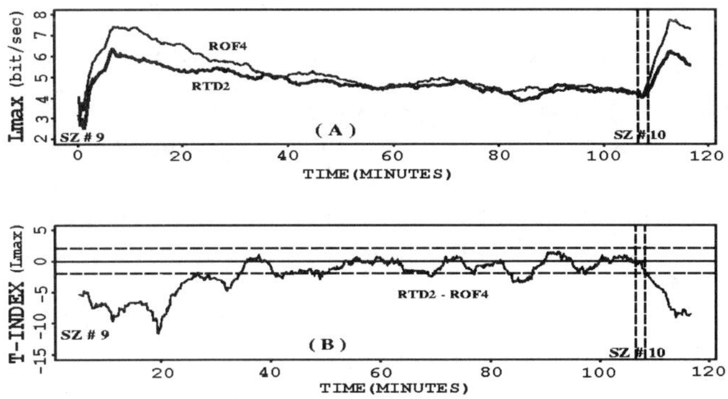


FIGURE 5. Preictal entrainment of a pair of brain sites prior to seizure ♯ 10. (a) $L_{max}$ values at the normal site ROF4 approach the ones at the epileptogenic site RTD2 about 33 minutes into the recording. Then, for 80 minutes up to seizure, the two sites interact with each other. (b) The $T$-index profile generated from the $L_{max}$ profiles in (a) is shown. The $\alpha = 0.05$ statistical significance level entrainment zone is also shown (two dashed horizontal lines). Entrainment of the two sites occurs when $T$-index values are within the depicted entrainment zone.

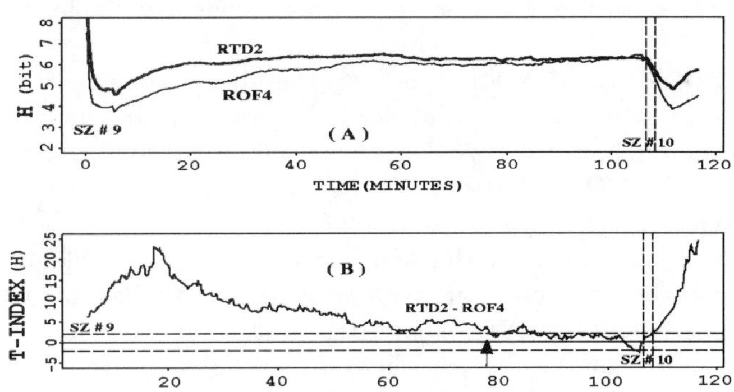


FIGURE 6. Preictal entrainment of a pair of brain sites prior to seizure ♯10. (a) $H$ values at the normal site ROF4 approach the ones at the epileptogenic site RTD2 about 30 minutes prior to seizure onset. (b) The $T$-index profile generated from the $H$ profiles in (a) is given. The $\alpha = 0.05$ statistical significance level entrainment zone is also shown (two dashed horizontal lines). Entrainment of the two sites occurs when $T$-index values are within the depicted entrainment zone.

critical sites, combined with the distance of the corresponding entropy ($H$) profiles. We will show the results of such an analysis for a series of seizures in one patient.

## 3. Optimization and spatiotemporal dynamical analysis

Quadratic binary programming has been extensively used to study Ising spin glass models. In these models the quadratic function is the Sherrington-Kirkpatric Hamiltonian that describes the mean-field theory of the spin glasses. This has motivated us to use quadratic binary programming to decide the selection of brain sites for optimal detection of the preictal transition [**ABW87, Bar82a, Bar82b, MPV87, HPT95**].

The general unconstrained quadratic zero-one programming problem (QP) has the form:

$$\text{(3.1)} \qquad \textit{global min } f(x) \;=\; c^T x + x^T Q x, \text{with } x \in \{0,\ 1\}^n$$

where $Q$ is an $n \times n$ symmetric matrix and $c$ is a vector of length $n$. Since $x_i^2 = x_i$ for $x_i \in \{0,\ 1\}$, $i = 1, \dots, n$ the linear part $c^T$ can be incorporated in the diagonal of $Q$.

An efficient branch and bound algorithm has been developed by Pardalos and Rodgers to solve general quadratic binary problems. The success of the algorithm is based on good lower bounds and a *dynamic preprocessing technique* that fixes variables during the branch and bound process.

The dynamic preprocessing technique is based on the *gradient sign* of an equivalent 0-1 quadratic problem (with zero diagonals in the corresponding matrix). If some partial derivative has constant sign, then the corresponding integer variable is known and can be fixed. This fixing of variables can also be applied to every subproblem of the branch and bound tree. Consequently, variables may be fixed dynamically.

The code Q01SUBS [**MW94**] (based on the above techniques) solves $0-1$ unconstrained quadratic problems both for dense and sparse matrices, as well as concave quadratic minimization problems with box constraints. The initial algorithm is described in [**PR89, PR90**].

If we consider the elements of matrix $\boldsymbol{Q}(n \times n)$ to be the $T_{ij}$ indices ($i = 1, ..., n; j = 1, ..., n$) that quantify the signed distances of brain sites $i$ and $j$ with respect to their estimated nonlinear dynamical measures (like entropy and Lyapunov exponents), then $\boldsymbol{Q}$ is an antisymmetric matrix. If the matrix $\boldsymbol{Q}$ is defined as $\boldsymbol{Q}_{ij} = \alpha_{ij} \times T_{ij} = (-\alpha_{ji}) \times (-T_{ji})$, where $\alpha_{ij}$'s are sign coefficients, $\boldsymbol{Q}$ becomes symmetric ($\boldsymbol{Q}_{ij} = \boldsymbol{Q}_{ji}$), and we can optimally select the sites $k$ such that $f(x)$ over these $k$ sites reaches global minimum $f(x^*)$. The $\alpha_{ij}$ can be defined as sign coefficients in any of the following four ways:

TABLE 1. The electrode sites chosen by the optimization program to minimize $f(x)$ are noted by 1. $T$-index values are estimated from the $STL_{max}$ profiles and $\alpha_{ij}$ from $H$ profiles as described in Case I, over a 10 minute preictal interval prior to each seizure.

| | Recording Areas | | | | | |
|---|---|---|---|---|---|---|
| Szs. | LTD | RTD | LST | RST | LOF | ROF |
| 1 | 11111 | 11111 | 1111 | 1111 | 1111 | 1111 |
| 2 | 11111 | 11111 | 1111 | 1111 | 1001 | 1011 |
| 3 | 11111 | 11111 | 1111 | 1111 | 1001 | 0011 |
| 4 | 11111 | 11111 | 1111 | 1111 | 1111 | 1111 |
| 5 | 11111 | 11111 | 1111 | 1111 | 1111 | 1111 |
| 6 | 11111 | 11111 | 1111 | 1111 | 1111 | 1111 |
| 7 | 11111 | 11111 | 1111 | 1111 | 1111 | 1111 |
| 8 | 11111 | 11111 | 1111 | 1111 | 1111 | 1111 |
| 9 | 11111 | 11111 | 1111 | 1111 | 1111 | 1111 |
| 10 | 11111 | 11111 | 1111 | 1111 | 1111 | 1111 |
| 11 | 11111 | 11111 | 1111 | 1111 | 1111 | 1111 |
| 12 | 11111 | 11111 | 1111 | 1111 | 1111 | 1111 |
| 13 | 11111 | 11111 | 1111 | 1111 | 1111 | 1111 |
| 14 | 11111 | 11111 | 1111 | 1111 | 1111 | 1111 |
| 15 | 11111 | 11111 | 1111 | 1100 | 1111 | 1111 |
| 16 | 11111 | 11111 | 1111 | 1100 | 1111 | 1111 |
| 17 | 11111 | 11111 | 1111 | 1111 | 1111 | 1111 |
| 18 | 11111 | 11111 | 1111 | 1111 | 1111 | 1111 |
| 19 | 11111 | 11111 | 1111 | 1111 | 1111 | 1111 |
| 20 | 11111 | 11111 | 1111 | 1111 | 1111 | 1111 |
| 21 | 11111 | 11111 | 1111 | 1111 | 1111 | 1111 |
| 22 | 11111 | 11111 | 1111 | 1111 | 1111 | 1111 |
| 23 | 11111 | 11111 | 1111 | 1111 | 1111 | 1111 |
| 24 | 11111 | 11111 | 1111 | 1111 | 1111 | 1111 |

Case I $\alpha_{ij} = sign(H_i - H_j)$ and $T_{ij}$ is the $T$-index of $L_{max}$ between site $i$ and $j$. Then $\boldsymbol{Q}_{ij}$ is negative (leading to minimum values of the objective function) if the entropy gradient is in the opposite direction from $L_{max}$ gradient (e.g., $\boldsymbol{Q}_{ij} < 0$ if: $\alpha_{ij} > 0$ and $T_{ij} < 0$, or $\alpha_{ij} < 0$ and $T_{ij} > 0$).

Case II $\alpha_{ij} = -sign(H_i - H_j)$ and $T_{ij}$ is the $T$-index of $L_{max}$ between site $i$ and $j$. Then $\boldsymbol{Q}_{ij}$ is negative if the entropy gradient is in the same direction as $L_{max}$ gradient.

Case III $\alpha_{ij} = sign(L_{max,i} - L_{max,j})$ and $T_{ij}$ is the $T$-index of $H$ between site $i$ and $j$. Then $\boldsymbol{Q}_{ij}$ is negative if the entropy gradient is in the opposite direction from $L_{max}$ gradient.

Case IV $\alpha_{ij} = -sign(L_{max,i} - L_{max,j})$ and $T_{ij}$ is the $T$-index of $H$ between site $i$ and $j$. Then $\boldsymbol{Q}_{ij}$ is negative if the entropy gradient is in the same direction as $L_{max}$ gradient.

TABLE 2. The electrode sites chosen by the optimization program to minimize $f(x)$ are noted by 1. $T$-index values are estimated from the $STL_{max}$ profiles and $\alpha_{ij}$ from $H$ profiles as described in Case II, over a 10 minute preictal interval prior to each seizure.

| Szs. | LTD | RTD | LST | RST | LOF | ROF |
|---|---|---|---|---|---|---|
| 1 | 11000 | 11000 | 0000 | 1100 | 0110 | 0100 |
| 2 | 00010 | 11001 | 0001 | 1001 | 0110 | 1100 |
| 3 | 00000 | 11001 | 1000 | 1101 | 1110 | 1111 |
| 4 | 00001 | 10001 | 0000 | 0101 | 0000 | 0000 |
| 5 | 00010 | 10010 | 0100 | 1001 | 0000 | 0000 |
| 6 | 00010 | 10101 | 0000 | 1001 | 1010 | 0100 |
| 7 | 00010 | 10000 | 1010 | 0000 | 0001 | 0000 |
| 8 | 00000 | 10001 | 0000 | 1100 | 0100 | 0000 |
| 9 | 00000 | 00000 | 1010 | 0001 | 0101 | 0000 |
| 10 | 00000 | 10101 | 0100 | 1100 | 0000 | 0011 |
| 11 | 00000 | 11001 | 1100 | 1000 | 0011 | 0011 |
| 12 | 00100 | 00000 | 0000 | 0011 | 0110 | 0100 |
| 13 | 00000 | 01000 | 0100 | 1000 | 0001 | 0000 |
| 14 | 00000 | 11000 | 0100 | 0000 | 0011 | 0000 |
| 15 | 00000 | 11101 | 0100 | 1011 | 0001 | 0011 |
| 16 | 11011 | 11111 | 1111 | 1011 | 1111 | 1111 |
| 17 | 00000 | 00000 | 1110 | 0010 | 1011 | 0000 |
| 18 | 00010 | 11100 | 1110 | 1000 | 1001 | 1001 |
| 19 | 00010 | 01000 | 0110 | 1000 | 1010 | 1000 |
| 20 | 00010 | 01000 | 1110 | 1000 | 1010 | 0000 |
| 21 | 00010 | 01000 | 1010 | 1001 | 1011 | 1000 |
| 22 | 00000 | 01000 | 1001 | 1011 | 1001 | 1000 |
| 23 | 00000 | 01000 | 1000 | 0011 | 0001 | 0000 |
| 24 | 00010 | 10100 | 0110 | 0100 | 0100 | 0000 |

TABLE 3. The electrode sites chosen by the optimization program to minimize $f(x)$ are noted by 1. $T$-index values are estimated from the $H$ profiles and $\alpha_{ij}$ from $STL_{max}$ profiles as described in Case III, over a 10 minute preictal interval prior to each seizure.

| Szs. | LTD | RTD | LST | RST | LOF | ROF |
|---|---|---|---|---|---|---|
| 1 | 11111 | 11111 | 1111 | 1111 | 1111 | 1111 |
| 2 | 11111 | 11111 | 1111 | 1111 | 1011 | 1111 |
| 3 | 11111 | 11111 | 1111 | 1111 | 1001 | 1011 |
| 4 | 11111 | 11111 | 1111 | 1111 | 1111 | 1111 |
| 5 | 11111 | 11111 | 1111 | 1111 | 1111 | 1111 |
| 6 | 11111 | 11111 | 1111 | 1111 | 1111 | 1111 |
| 7 | 11111 | 11111 | 1111 | 1111 | 1111 | 1111 |
| 8 | 11111 | 11111 | 1111 | 1111 | 1111 | 1111 |
| 9 | 11111 | 11111 | 1111 | 1111 | 1111 | 1111 |
| 11 | 11111 | 11111 | 1111 | 1111 | 1111 | 1111 |
| 11 | 11111 | 11111 | 1111 | 1111 | 1111 | 1111 |
| 12 | 11111 | 11111 | 1111 | 1111 | 1111 | 1111 |
| 13 | 11111 | 11111 | 1111 | 1111 | 1111 | 1111 |
| 14 | 11111 | 11111 | 1111 | 1111 | 1111 | 1111 |
| 15 | 11111 | 11111 | 1111 | 1100 | 1111 | 1111 |
| 16 | 11111 | 11111 | 1111 | 1110 | 1111 | 1111 |
| 17 | 11111 | 11111 | 1111 | 1111 | 1111 | 1111 |
| 18 | 11111 | 11111 | 1111 | 1111 | 1111 | 1111 |
| 19 | 11111 | 11111 | 1111 | 1111 | 1111 | 1111 |
| 21 | 11111 | 11111 | 1111 | 1111 | 1111 | 1111 |
| 21 | 11111 | 11111 | 1111 | 1111 | 1111 | 1111 |
| 22 | 11111 | 11111 | 1111 | 1111 | 1111 | 1111 |
| 23 | 11111 | 11111 | 1111 | 1111 | 1111 | 1111 |
| 24 | 11111 | 11111 | 1111 | 1111 | 1111 | 1111 |

TABLE 4. The electrode sites chosen by the optimization program to minimize $f(x)$ are noted by 1. $T$-index values are estimated from the $H$ profiles and $\alpha_{ij}$ from $STL_{max}$ profiles as described in Case IV, over a 10 minute preictal interval prior to each seizure.

| | Recording Areas | | | | | |
|---|---|---|---|---|---|---|
| Szs. | LTD | RTD | LST | RST | LOF | ROF |
| 1 | 11000 | 11000 | 0000 | 1100 | 0100 | 0100 |
| 2 | 00000 | 00111 | 1000 | 0000 | 1110 | 1100 |
| 3 | 00000 | 11001 | 1000 | 1101 | 1110 | 1100 |
| 4 | 00000 | 00101 | 0110 | 0011 | 0000 | 0000 |
| 5 | 00110 | 01001 | 0000 | 1110 | 0001 | 0000 |
| 6 | 00000 | 10101 | 0000 | 1001 | 1010 | 0100 |
| 7 | 10000 | 00000 | 0001 | 0000 | 1010 | 1100 |
| 8 | 00000 | 10001 | 0100 | 0100 | 0101 | 0000 |
| 9 | 00010 | 00000 | 0000 | 0011 | 1010 | 1100 |
| 10 | 00000 | 10001 | 1100 | 1100 | 0100 | 0001 |
| 11 | 00000 | 11000 | 1100 | 0000 | 0011 | 0011 |
| 12 | 00000 | 00000 | 1001 | 0001 | 1100 | 1100 |
| 13 | 00000 | 01000 | 0100 | 0000 | 0011 | 0000 |
| 14 | 00000 | 01000 | 0100 | 0000 | 0001 | 0100 |
| 15 | 00001 | 11101 | 0100 | 1011 | 0101 | 0011 |
| 16 | 11010 | 11111 | 1111 | 1111 | 1111 | 1111 |
| 17 | 00000 | 11111 | 0000 | 1000 | 0100 | 0011 |
| 18 | 00010 | 11100 | 1110 | 1000 | 1001 | 1001 |
| 19 | 00010 | 01000 | 1110 | 0000 | 1010 | 1000 |
| 20 | 00010 | 01000 | 1110 | 1000 | 1010 | 0000 |
| 21 | 00010 | 01000 | 1010 | 1001 | 1011 | 1000 |
| 22 | 00000 | 01000 | 1001 | 1011 | 1001 | 1000 |
| 23 | 00000 | 01001 | 0100 | 0001 | 0101 | 0000 |
| 24 | 00000 | 00000 | 1000 | 0010 | 1011 | 1000 |

## 4. Results

### 4a. Detection of the preictal transition

In this section, results from the application of the above described optimization method to the selection of brain sites for optimal detection of the preictal transition in epileptic seizures are shown. The method is applied to 24 periods, each of 10 minutes in duration, prior to corresponding epileptic seizures, in one patient with a right hippocampal epileptogenic focus (RTD). Selection of sites from the global minimization of the matrix $\boldsymbol{Q}$ in cases I through IV are respectively shown in Tables I through IV. A careful examination of the sites selected from the application of quadratic integer programming reveals that: in cases I and III almost every available recording site is selected throughout all seizures (see Tables I and III). On the contrary, in cases II and IV, a substantially smaller subset of sites is selected (see Tables II and IV). In addition, the global minima $f(x^*)$ in cases I and III, are much smaller than the ones in cases II and IV (see Figures 7 and 8). Both these observations imply that only cases II and IV optimally characterize the transition to epileptic seizures. This means that, in the epileptic transition, the overall direction

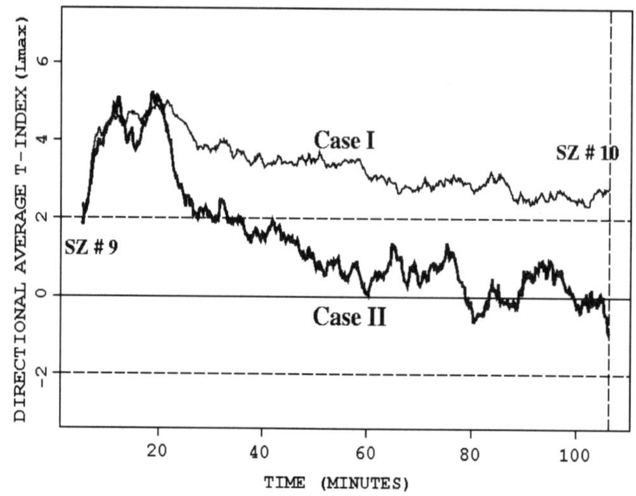


FIGURE 7. Global minimum $f(x^*) = \sum_{ij} \boldsymbol{Q}_{i,j}$ over time of the distances between dynamical measures $STL_{max}$ over all sites (regular line - Case I) and optimally selected sites (bold line - case II) from the 10 minute preictal period of seizure ♯ 10.

of the entropy gradient over selected sites should be the same as the direction of the $STL_{max}$ gradient over these sites. As we will see in the next subsection, brain sites that mostly satisfy this behavior preictally belong to the epileptogenic zone.

Comparing Figure 7 and Figure 8, we also notice that sites selected preictally from the global minimization of the $T$-index of $STL_{max}$ (case II) converge much earlier than the ones selected from the global minimization of the $T$-index of $H$ (case IV). This implies that, preictally, the entrainment of $STL_{max}$ profiles precedes the entrainment of entropy profiles and can be used as an early warning for an impending epileptic seizure.

### 4b. Epileptogenic focus localization

The degree of participation of individual brain sites in the preictal transition of the 24 seizures we analyzed is given in Tables V and VI. The degree of participation of a site is measured by the number of seizures this site participates in, as selected by the optimization algorithm. In Table V, degree of participation entries were estimated from Table II ($T_{ij}$ from $STL_{max}$ measures). In Table VI, entries were estimated from Table IV ($T_{ij}$ from $H$ measures). From Table V we notice that RTD1, RTD2 and RST1 exhibit the maximum degree of participation in the preictal entrainment in this patient. From Table VI we notice that RTD2 and RST1 exhibit the maximum participation. The brain site RTD2, selected in both Tables V and VI, is the site that exhibits first the seizure onset (see Figure 2) in all 24 seizures, and as such, according to the current golden rule in epileptology, it qualifies as the most pathologic site. Hence, by optimal selection of brain sites that are entrained

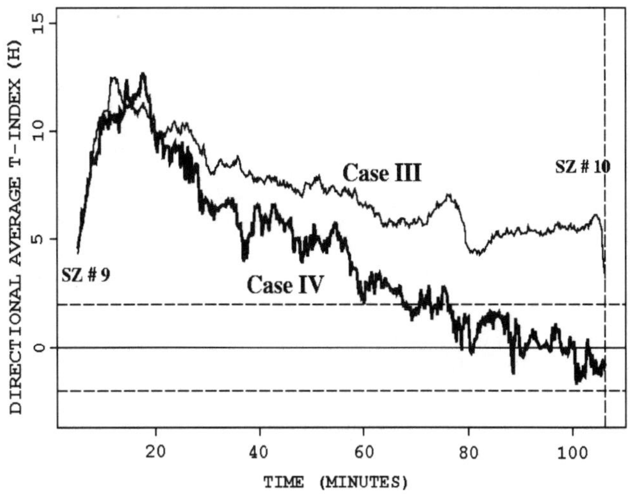


FIGURE 8. Global minimum $f(x^*) = \sum_{ij} \boldsymbol{Q}_{i,j}$ over time of the distances between dynamical measures $H$ over all sites (regular line - Case III) and optimally selected sites (bold line - case IV) from the 10 minute preictal period of seizure ♯ 10.

TABLE 5. Degree of participation of selected sites in the preictal transition of 24 seizures ($T_{ij}$ from $STL_{max}$ - Case II).

| L Sites | No. Szs. | R Sites | No. Szs. | (L - R) |
|---|---|---|---|---|
| LTD1 | 2 | RTD1 | 15 | -13 |
| LTD2 | 2 | RTD2 | 14 | -12 |
| LTD3 | 1 | RTD3 | 6 | -5 |
| LTD4 | 10 | RTD4 | 2 | +7 |
| LTD5 | 2 | RTD5 | 9 | -7 |
| LST1 | 11 | RST1 | 16 | -5 |
| LST2 | 12 | RST2 | 6 | +6 |
| LST3 | 9 | RST3 | 6 | +3 |
| LST4 | 3 | RST4 | 12 | -9 |
| LOF1 | 9 | ROF1 | 7 | +2 |
| LOF2 | 8 | ROF2 | 6 | +2 |
| LOF3 | 12 | ROF3 | 5 | +7 |
| LOF4 | 12 | ROF4 | 6 | +6 |
| Total (L) | 93 | (R) | 110 | -17 |

in the preictal transition we are able to identify epileptogenic focal sites in this patient.

The differences between degrees of participation of homologous sites of the left and right brain hemispheres are given in the last columns of Tables V and VI. The

TABLE 6. Degree of participation of selected sites in the preictal transition of 24 seizures ($T_{ij}$ from $H$ - Case IV).

| L Sites | No. Szs. | R Sites | No. Szs. | (L - R) |
|---|---|---|---|---|
| LTD1 | 3 | RTD1 | 10 | -7 |
| LTD2 | 2 | RTD2 | 15 | -13 |
| LTD3 | 1 | RTD3 | 7 | -6 |
| LTD4 | 7 | RTD4 | 3 | +4 |
| LTD5 | 1 | RTD5 | 11 | -10 |
| LST1 | 12 | RST1 | 12 | 0 |
| LST2 | 12 | RST2 | 6 | +6 |
| LST3 | 6 | RST3 | 7 | -1 |
| LST4 | 4 | RST4 | 10 | -6 |
| LOF1 | 13 | ROF1 | 11 | +2 |
| LOF2 | 10 | ROF2 | 9 | +1 |
| LOF3 | 12 | ROF3 | 4 | +8 |
| LOF4 | 12 | ROF4 | 6 | +6 |
| Total (L) | 95 | (R) | 111 | -16 |

maximum absolute |L-R| differences are noted between sites from LTD and RTD areas. This means that the epileptogenic RTD area participates in the unfolding of a seizure more than its homologous contralateral "normal" LTD area in the left hemisphere. In particular, in Table V, RTD1 and RTD2 exhibit the maximum |L-R| differences, denoting a mesial focus in the right hippocampus. The corresponding selection in Table VI is RTD2 and RTD5, denoting a more diffuse localization of the focus in the right hippocampus by entropies than by $L_{max}$. These differences may be explained in terms of spatiotemporal dynamics: since, as we have shown in the previous section, $H$ evolves slower than $STL_{max}$. Thus, one should expect a more diffuse spatial selection of sites when $T_{ij}$ is estimated from $H$ than from $STL_{max}$.

## 5. Conclusions

We have used quadratic integer programming techniques, for the optimal selection of communicating brain sites at periods prior to epileptic seizures. The objective function to be minimized was the distance of chaoticity (entropy rate) and complexity (entropy) measures between recording brain sites. These measures were estimated by a nonlinear dynamical analysis of the recorded electrical activity (EEG) of the brain at each site. The sites selected by the optimization method have provided two important insights. First, the degree of participation of each site in the preictal transition can be used as an index of how pathologic this site is, and may eventually lead to the identification of the epileptogenic zone (see Tables 5 and 6). Second, following the dynamical measures of the thus optimally selected sites over time we can detect the preictal transition long before (terms of minutes) the occurrence of a seizure and may thus predict an epileptic seizure. Inclusion of all recording sites does not lead to prediction of seizures (see Figures 7 and 8). Third, selection of sites based on $L_{max}$ measures results to earlier detection of the preictal transition than the one based on entropy measures. Ongoing computational studies on other patients will be reported in the future.

## References

[Aba96] H. D. I. Abarbanel, *Analysis of observed chaotic data*, Springer-Verlag, New York, 1996.

[ABW87] G. G. Athanasiou, C. P. Bachas, and W. F. Wolf, *Invariant Geometry of Spin-glass States*, Phy. Rev. B **35** (1987), 1965–1968.

[ASSG87] B. W. Abou-Khalil, G. J. Seigel, J. C. Sackellares, S. Gilmore, R. Hichwa, and R. Marshall, *Positron emission tomograghy studies of cerebral glucose metabolism in patients with chronic partial epilepsy*, Ann. Neurol. **22** (1987), 480–486.

[Bar82a] F. Barahona, *On the computational complexity of spin glass models*, J. Phys. A: Math. Gen. **15** (1982), 3241–3253.

[Bar82b] F. Barahona, *On the exact ground states of three-dimensional ising spin glasses*, J. Phys. A: Math. Gen. **15** (1982), L611–L615.

[BD86] A. Babloyantz and A. Destexhe, *Low dimensional chaos in an instance of epilepsy*, Proc. Natl. Acad. Sci. USA **83** (1986), 3513–3517.

[Ber29] H. Berger, *Uber das elektroenkephalogramm des menchen*, Arch. Psychiatr. Nervenkr. **87** (1929), 527–570.

[BSHR95] D. E. Burdette, Sakuraisy, T. R. Henry, D. A. Ross, P. B. Pennell, K. A. Frey, J. C. Sackellares, and R. Albin, *Temporal lobe central benzodiazepine binding in unilateral mesial temporal lobe epilepsy*, Neurology **45** (1995), 934–941.

[Cat75] R. Caton, *The electric currents of the brain*, BMJ **2** (1875), 278.

[CB90] G. Casella and R. L. Berger, *Statistical Inference*, Duxbury Press, Belmont, CA, 1990.

[CISR96] M. Casdagli, L. D. Iasemidis, J. C. Sackellares, S. N. Roper, R. L. Gilmore, and R. S. Savit, *Characterizing nonlinearity in invasive EEG recordings from temporal lobe epilepsy*, Physica D **99** (1996), 381–399.

[CIGR97] M. Casdagli, L. D. Iasemidis, R. L. Gilmore, S. N. Roper, R. S. Savit, and J. C. Sackellares, *Nonlinearity in invasive EEG recordings from patients with temporal lobe epilepsy*, Electroenceph. Clin. Neurophysiol. **102** (1997), 98–105.

[EKRC86] J. P. Eckmann, S. O. Kamphorst, D. Ruelle, and S. Ciliberto, *Lyapunov exponents from time series*, Phys. Rev. A **34** (1986), 4971–4972.

[EL98] C. E. Elger and K. Lehnertz, *Seizure prediction by non-linear time series analysis of brain electrical activity*, Europ. J. Neurosci., **10** (1998), 786–789.

[ERKS94] T. Elbert, W. J. Ray, J. Kowalik, J. E. Skinner, K. E. Graf, and N. Birbaumer, *Chaos and physiology: Deterministic chaos in excitable cell assemblies*, Physiol. Rev. **74** (1994), 1–47.

[EKPM82] J. Engel Jr., D. E. Kuhl, M. E. Phelps, and J. C. Mazziota, *Interictal cerebral glucose metabolism in partial epilepsy and its relation to EEG changes*, Ann. Neurol. **12** (1982), 510–517.

[FMWB99] M. Feucht, U. Moller, H. Witte, F. Benninger, S. Asenbaum, D. Prayer, and M. H. Friedrich, *Application of correlation dimension and pointwise dimension for non-linear topographical analysis of focal onset seizures*, Med. Biol. Eng. Comp., **37** (1999), 208–217.

[FSC64] M. A. Falconer, E. A. Serefetinides, and J. A. N. Corsellis, *Aetiology and pathogenesis of temporal lobe epilepsy*, Arch. Neurol. **19** (1964), 233–240.

[FS86] A. M. Fraser and H. L. Swinney, *Independent coordinates for strange attractors from mutual information*, Phys. Rev. A **33** (1986), 1134–1140.

[Glo69] P. Gloor, *Hans Berger on the electroencephalogram of man*, Elsevier, Amsterdam, 1969.

[GP83a] P. Grassberger and I. Procaccia, *Measuring the strangeness of strange attractors*, Physica D **9** (1983), 189–208.

[GP83b] P. Grassberger and I. Procaccia, *Characterization of strange attractors*, Phys. Rev. Lett. **50** (1983), 346–349.

[GSS91] P. Grassberger, T. Schreiber, and C. Schaffrath, *Nonlinear time sequence analysis*, Int. J. Bifurc. Chaos **1** (1991), 521–547.

[Hol86] A. V. Holden, *Chaos-nonlinear science: theory and applications*, Manchester University Press, Manchester, 1986.

[HPT95] H. Horst, P. M. Pardalos , and V. Thoai, *Introduction to global optimization, Series on Nonconvex Optimization and its Applications*, **3**, Kluwer Academic Publishers, Dordrecht, 1995.

[Ias91] L. D. Iasemidis, *On the dynamics of the human brain in temporal lobe epilepsy*, Ph.D. Dissertation, University of Michigan, Ann Arbor, 1991.

[IOSS94] L. D. Iasemidis, L. D. Olson, J. C. Sackellares, and R. Savit, *Time dependencies in the occurrences of epileptic seizures: a nonlinear approach*, Epilepsy Research **17** (1994), 81–94.

[IPCG97] L. D. Iasemidis, J. C. Principe, J. M. Czaplewski, R. L. Gilmore, S. N. Roper, and J. C. Sackellares, *Spatiotemporal transition to epileptic seizures: A non linear dynamical analysis of scalp and intracranial EEG recordings*, Spatiotemporal Models in Biological and Artifical Systems (Amsterdam)(F. Lopes da Silva, J. C. Principe, and L. B. Almeida, eds.), IOS Press, 1997, pp. 81–88.

[IPS96] L. D. Iasemidis, J. C. Principe, and J. C. Sackellares, *Spatiotemporal dynamics of human epileptic seizures*, 3rd Experimental Chaos Conference (Singapore)( R. G. Harrison, W. Lu, W. Ditto, L. Pecora, M. Spano, and S. Vohra, eds.), World Scientific, 1996, pp. 26–30.

[IPS00] L. D. Iasemidis, J. C. Principe, and J. C. Sackellares, *Measurement and quantification of spatiotemporal dynamics of human epileptic seizures*, Nonlinear biomedical signal processing, **2** (M. Akay, ed.), IEEE Press, in press.

[IS91] L. D. Iasemidis and J. C. Sackellares, *The temporal evolution of the largest Lyapunov exponent on the human epileptic cortex*, Measuring chaos in the human brain (Singapore)(D. W. Duke and W. S. Pritchard, eds.), World Scientific, 1991, pp. 49-82.

[IS96] L. D. Iasemidis and J. C. Sackellares, *Chaos theory and epilepsy*, The Neuroscientist **2** (1996), 118–126.

[ISS93] L. D. Iasemidis, J. C. Sackellares, and R. S. Savit, *Quantification of hidden time dependencies in the EEG within the framework of nonlinear dynamics*, Nonlinear dynamical analysis of the EEG (Singapore)(B. H. Jansen and M. E. Brandt, eds.), World Scientific, 1993, pp. 30–47.

[ISZW90] L. D. Iasemidis, J. C. Sackellares, H. P. Zaveri, and W. J. Williams, *Phase space topography of the electrocorticogram and the Lyapunov exponent in partial seizures*, Brain Topogr. **2** (1990), 187–201.

[IZSW88] L. D. Iasemidis, H. P. Zaveri, J. C. Sackellares, and W. J. Williams, *Linear and nonlinear modeling of ECoG in temporal lobe epilepsy*, 25th Annual Rocky Mountain Bioengineering Symposium **24** (1988), 187–193.

[Jan91] B. H. Jansen, *Is it and so what? A critical review of EEG-chaos*, Measuring chaos in the human brain (Singapore)(D. W. Duke and W. S. Pritchard, eds.), World Scientific, 1991, pp. 83–96.

[Kol54] A. N. Kolmogorov, *The general theory of dynamical systems and classical mechanics*, Foundations of Mechanics (R. Abraham and J. E. Marsden, eds.), 1954.

[Kos92] E. J. Kostelich, *Problems in estimating dynamics from data*, Physica D **58** (1992), 138–152.

[LE95] K. Lehnertz, and C. E. Elger, *Spatio-temporal dynamics of the primary epileptogenic area in temporal lobe epilepsy characterized by neuronal complexity loss*, Electroenceph. Clin. Neurophysiol, **95** (1995), 108–117.

[LE98] K. Lehnertz, and C. E. Elger, *Can epileptic seizures be predicted? Evidence from nonlinear time series analysis of brain electrical activity*, Phys. Rev. Lett., **80** (1998), 5019–5022.

[Lop87] F. Lopes da Silva, *EEG analysis: theory and practice; Computer-assisted EEG diagnosis: Pattern recognition techniques*, Electroencephalography: Basic principles, clinical applications and related field (Baltimore)(E. Niedermeyer and F. Lopes da Silva (eds.), Urban and Schwarzenberg, 1987, pp. 871–897.

[LMAV00] M. Le Van Quyen, J. Martinerie, C. Adam, and F. J. Varela, *Nonlinear spatio-temporal interdependences of interictal intracranial EEG recordings from patients with temporal lobe epilepsy: localizing of epileptogenic foci*, Physica D, in press.

[LMBV99] M. Le Van Quyen, J. Martinerie, M. Baulac, and F. Varela, *Anticipating epileptic seizures in real time by a non-linear analysis of similarity between EEG recordings*, NeuroReport, **10** (1999), 2149–2155.

[MALB98] J. Martinerie, C. Adam, M. Le Van Quyen, M. Baulac, S. Clemenceau, B. Renault, and F. J. Varela, *Epileptic seizures can be anticipated by non-linear analysis*, Nature Medicine, **4** (1998), 1173–1176.

[MC66] J. H. Margerison and J. A. N. Corsellis, *Epilepsy and the temporal lobes*, Brain, **89** (1966), 499–530.

[May86] G. Mayer-Kress, *Dimension and entropies in chaotic systems*, Springer-Verlag, Berlin, 1986.

[MGHS91] J. W. McDonald, E. A. Garofalo, T. Hood, J. C. Sackellares, S. Gilmore, P. E. McKeever, J. C. Troncaso and M. V. Johnston, *Altered excitatory and inhibitory aminoacid receptor binding in hippocampus of patients with temporal lobe epilepsy*, Annals of Neurology, **29** (1991), 529–541.

[MPV87] M. Mezard, G. Parisi, and M. A. Virasoro, *Spin Glass Theory and Beyond*, World Scientific, Singapore, 1987.

[MW94] J. J. Moré and S. J. Wright, *Optimization Software Guide*, SIAM, Philadelphia, 1994.

[Nie87] E. Niedermeyer, *Depth Electroencephalography*, Electroencephalography: Basic principles, clinical applications and related fields (Urban and Schwarzenberg, Baltimore)( E. Niedermeyer and F. Lopes da Silva, eds.), 1987, pp. 593–617.

[Ose68] A. Oseledec, *A multiplicative ergodic theorum-Lyapunov characteristic numbers for dynamical systems (English translation)*, IEEE Int. Conf. ASSP, **19** (1968), 179–210.

[PCFS80] N. H. Packard, J. P. Crutchfield, J. D. Farmer, and R. S. Shaw, *Geometry from time series*, Phys. Rev. Lett., **45** (1980), 712–716.

[PAD93] M. Palus, V. Albrecht, and I. Dvorak, *Information theoretic test for nonlinearity in time series*, Phys. Lett. A, **175** (1993), 203–209.

[Pes77] J. Pesin, *Characteristic Lyapunov exponents and smooth ergodic theory*, Russian Math. Survey, **4** (1977), 55–114.

[PR89] P. M. Pardalos and G. Rodgers, *Parallel Branch and Bound Algorithms for Unconstrained Quadratic Zero-one Programming*, Impact of Recent Computer Advances on Operations Research (R. Sharda et al., eds.), North-Holland, 1989, pp. 131–143.

[PR90] P. M. Pardalos and G. Rodgers, *Computational aspects of a branch and bound algorithm for quadratic zero-one programming*, Computing, **45** (1990), 131–144.

[Ren70] A. Renyi, *Probability theory*, Elsevier, Amsterdam, 1970.

[SIGR00] J. C. Sackellares, L. D. Iasemidis, R. L. Gilmore, and S. N. Roper, *Epilepsy - when chaos fails*, Chaos in the brain? (Singapore) (P. Grassberger, C. E. Elger, and K. Lehnertz, eds.), World Scientific, in press.

[SIZW89a] J. C. Sackellares, L. D. Iasemidis, H. P. Zaveri, W. J. Williams, and T. W. Hood, *Inference on the chaotic behavior of the epileptogenic focus*, Epilepsia, **29** (1989), 682.

[SIZW89b] J. C. Sackellares, L. D. Iasemidis, H. P. Zaveri, and W. J. Williams, *Measurement of chaos to localize seizure onset*, Epilepsia, **30** (1989), 663.

[SSAH90] J. C. Sackellares, G. J. Siegel, B. W. Abou-Khalil, T. W. Hood, S. Gilmore, P. McKeever, R. D. Hichwa, and G. D. Hutchins, *Differences between lateral and mesial temporal metabolism interictally in epilepsy of mesial temporal origin*, Neurology, **40** (1990), 1420–1426.

[Tak81] F. Takens, *Detecting strange attractors in turbulence*, Dynamical systems and turbulence, Lecture notes in mathematics (Heidelburg)(D. A. Rand and L. S. Young, eds.), Springer-Verlag, 1981, pp. 366–381.

[VK86] J. A. Vastano and E. J. Kostelich, *Comparison of algorithms for determining Lyapunov exponents from experimental data*, Dimensions and entropies in chaotic systems: quantification of complex behavior (Berlin)(G. Mayer-Kress , ed.), Springer-Verlag, 1986, pp. 100–107.

[VVHP99] M. J. Van der Heyden, D. N. Velis, B. P. T. Hoekstra, J. P. M. Pijn, W. Van Emde Boas, C. W. M. Van Veelen, P. C. Van Rijen, F. H. Lopes da Silva, and J. DeGoede, *Non-linear analysis of intracranial human EEG in temporal lobe epilepsy*, Clinical Neurophysiology, **110** (1999), 1726–1740.

[Wal82] P. Walters, *An intorduction to ergodic theory*, Springer-Verlag, Berlin, 1982.

[WSSV85] A. Wolf, J. B. Swift, H. L. Swinney, and J. A. Vastano, *Determining Lyapunov exponents from a time series*, Physica D, **16** (1985), 285–317.

[Wei90] Weisberg, *Applied linear regression*, Wiley, New York, 1990.

Electrical and Computer Engineering and Neuroscience, University of Florida
*E-mail address*: `leon@epilepsy.health.ufl.edu`

Statistics, University of Florida
*E-mail address*: `shiau@epilepsy.health.ufl.edu`

Neurology, Bioengineering, Neuroscience, University of Florida
*E-mail address*: `sackellares@epilepsy.health.ufl.edu`

Industrial and Systems Engineering, University of Florida
*E-mail address*: `pardalos@ufl.edu`

DIMACS Series in Discrete Mathematics
and Theoretical Computer Science
Volume **55**, 2000

# Analysis of Magnetic Resonance Angiography Images Using Skeletonization and Distance Transforms

Ingela Nyström and Örjan Smedby

ABSTRACT. In magnetic resonance angiography (MRA), large 3D data sets of the vascular system are acquired. This paper deals with two techniques to represent relevant information from such data sets in a more compact format. After segmentation of the vessels, the curve skeleton is extracted by an algorithm based on the distance transformation. Using the $D^6$ distance to the background, the algorithm first reduces the original object to a surface skeleton and then to a curve skeleton, after which "pruning" can be performed to remove irrelevant small branches. Applying this procedure to MRA data from the pelvic arteries resulted in a good description of the tree structure of the vessels with a much smaller number of voxels. To detect stenoses, 2D projections such as maximum intensity projection (MIP) are usually employed, but these often fail to demonstrate a stenosis if the projection angle is not suitably chosen. A new method combines the distance-labelled curve skeleton with the reverse distance transformation to visualize stenoses independent of viewing direction. Experiments with synthetic data indicate that stenoses invisible in an ordinary projection may be seen with this technique.

## 1. Introduction

Both magnetic resonance angiography (MRA) [**AET93, Pri94**] and computed tomography angiography (CTA) [**NMR$^+$92**] are methods that can produce (3D) volume images of the vasculature in the human body. MRA, which has been used here, is a magnetic resonance imaging (MRI) method to assess vascular disease that has spread widely into the clinical routine.

Volume imaging usually implies large amounts of data, and it is, in general, a non-trivial problem how to present these volumes on 2D devices such as screens and photographic film. When interpreting and analysing MRA images, results are usually presented as maximum intensity projections (MIP), multi-planar reconstructions (MPR), or volume renderings [**DHW$^+$97**]. The choice of viewing direction for these (and other) projections is of great importance for the detection of stenoses.

---

2000 *Mathematics Subject Classification.* Primary 68T10, 92C50; Secondary 57N99, 52C99.
*Key words and phrases.* pattern recognition, shape representation, 3D skeleton, reverse distance transform, MR angiography, arteries, stenosis detection.

Despite the large data volumes, the relevant information in an MRA or CTA data set is rather limited. Usually, the diagnostic physician is interested in the vessel anatomy, i.e., in the tree structure of the arteries, and in variations in vessel calibre. If one can identify the vessels with their branching points and the stenoses or narrow parts, this is usually an adequate description of the vessel system. In this paper, we present a method to use a distance-labelled *3D curve skeleton* to extract this important information from a volume image of the vascular system. We will also introduce a new presentation method aiming at visualizing variations in vessel width independent of viewing direction.

Skeletonization, also referred to as thinning, is a process where objects are reduced to structures of lower dimension while preserving topological properties as well as the general shape of the object. Two main classes of skeletonization methods for 2D images have been developed: *iterative thinning* and *distance transform analysis*. In 3D, the published methods have almost exclusively been based on iterative thinning, e.g., [**Ber95**] and [**SCM97**]. Some published methods produce surface skeletons, some of them reduce the objects to curve skeletons. Thin elongated objects, such as blood vessels, are well suited for reduction to curve skeletons. Earlier work with the aim to obtain curve skeletons for applications can be found, e.g., in [**MS96**], in [**GSV96**], where the skeleton concept from [**LKC94**] is adopted, and in [**PK99**]. The methods were applied to visual analysis of computed tomography (CT) lung data, interactive virtual colonoscopy rendered from spiral CT scans, and ventricle extraction from MRI brain data, respectively.

The method used in this paper for reducing objects in binary volume images to curve skeletons is described in detail in [**BNSdB99**]. It is a skeletonization based on iterative thinning, but guided by a distance transform. After skeletonization, the blood vessels consist of thread-like structures, which are one voxel thick and located centrally in the vessel lumen, while containing the distance to the closest background voxel from every voxel. These distances are utilized in our presentation method. For example, a stenosis reducing the anteroposterior diameter of a craniocaudal vessel, which is normally seen only in the lateral projection, will then become visible also in the frontal projection.

This work is a continuation of the work presented in [**Nys98**]. Here, we show a useful application for our earlier work. MRA is briefly introduced in Section 2. Distance transforms are briefly described in Section 3.1, including a method for reversing the skeleton, while the skeletonization method is outlined in Section 3.2. Resulting skeletons will be shown for synthetic and MRA images in Section 4. In Section 5 we show how the curve skeletons may be utilized in a presentation method.

## 2. Magnetic resonance angiography

Magnetic resonance angiography is a technique, or rather a set of techniques, to non-invasively image the flowing blood in vessels. The physical principles can vary, and there are three methods in common clinical use today:

**inflow angiography:** also known as time of flight (TOF), which uses flow effects on the signal amplitude,

**phase contrast angiography:** which uses flow effects on the phase of the signal, and

**Gadolinium angiography:** which does not rely on flow effects, but on the effect on MR signal by a paramagnetic contrast material, Gadolinium (Gd), which is given as an intravenous injection in order to enhance the arteries.

In all three cases, the most common situation involves acquiring a 3D data set, and then presenting the data on a screen, either as a single 2D slice (which may have a different orientation from the originally acquired slices) or using rendering algorithms such as the most commonly used maximum intensity projection (MIP). An important issue with all projection methods is the choice of viewing direction.

In our study, Gd-MRA images were taken from the pelvic arteries in a patient with atherosclerosis. Figure 1 (top) shows one slice in the data set, showing only parts of the vessel system. Figure 1 (bottom), presents the volume as a MIP, clearly showing the vessel system, but with no depth information in it.

## 3. Image processing methods

Our method employs two central concepts from image processing: *skeletonization* and (ordinary and reverse) *distance transformation*. It should be noted that here we assume that binary images have been obtained, e.g., by segmentation into object and background.

**3.1. Distance transformation.** Each object voxel can be labelled with (an approximation of) the *distance* to the closest background voxel. In general, the best metric to choose when computing the distance depends on the application. The three simplest metrics for volume images are the "path-generated" distances $D^6$, $D^{18}$, and $D^{26}$, where the distance is defined as the number of steps in the minimal path between voxels with 6-, 18-, and 26-connectedness, respectively [**Bor96**]. In this paper we will use only $D^6$, i.e., the 3D equivalent of the city-block distance, where steps only to the six face-neighbours of a voxel are allowed. The $D^6$ distance transform of the object consists of values representing the number of steps to the closest background voxel. These values are ordered in layers that you may imagine can be peeled off one value at a time. All voxels with a face-neighbour in the background will then have the same distance value. To compute the distances in an image a *distance transformation* (DT) is used, here $DT^6$. A distance transformation relies on the same principle in the 3D case as in the 2D case [**Bor84**]. The $DT^6$ is computed by a forward and a backward scan through the image, propagating increasing distances. Each voxel is assigned the minimum of the value of the current voxel and the values of the already visited face-neighbours added by one.

The distance label of any voxel in the distance transform can be interpreted as the radius of a digital ball — a solid sphere — centred on that voxel. Balls look different for different metrics, e.g., the $D^6$ ball is an octahedron. The definition of the distance transform implies that each ball is as large as possible still included in the object.

The *reverse* distance transformation starts from a number of seed voxels with distance values [**RP66**]. Each distance value corresponds to the radius of a ball. The result is an object with the shape of the union of the balls associated to the distance values. The reverse DT can also (as the ordinary DT) be computed by two raster scans, but this time decreasing distances are propagated. Each voxel is assigned the maximum of the value of the current voxel and the values of the

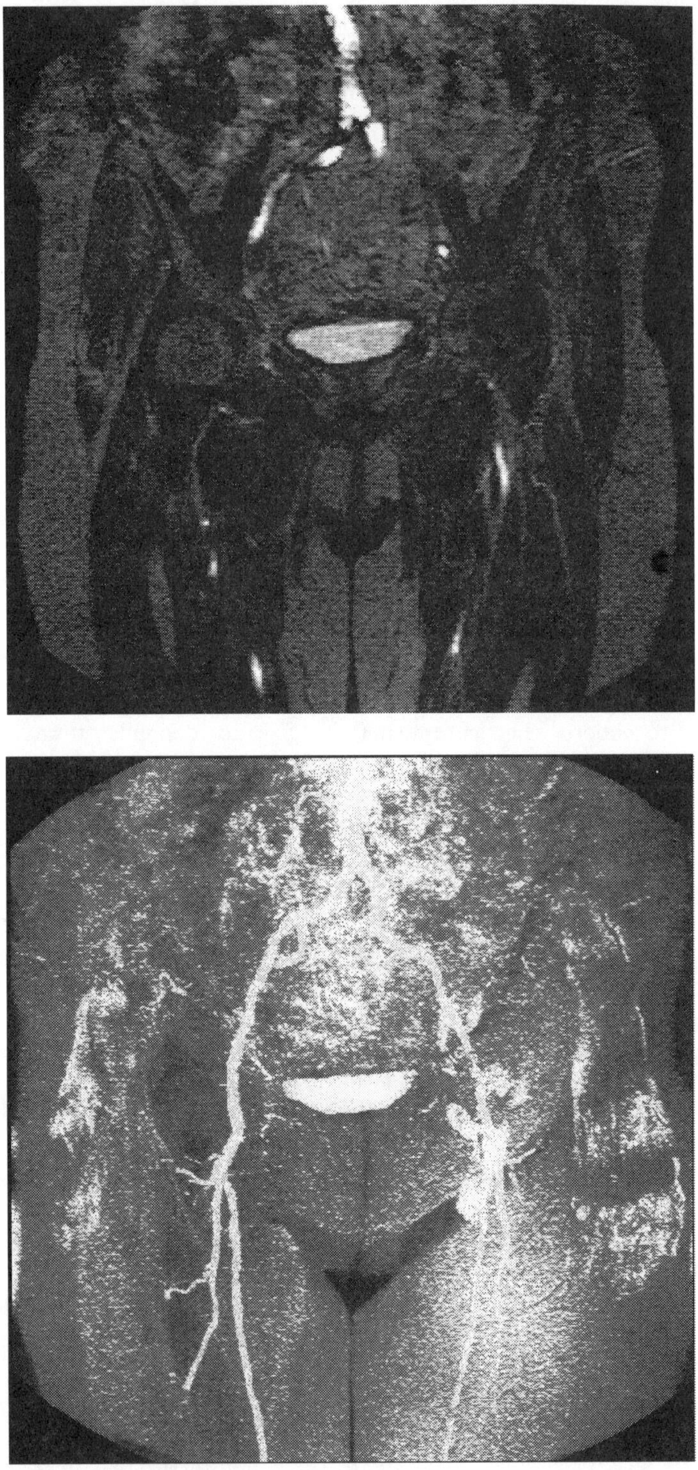

FIGURE 1. Two different presentations of an MRA image. Top: sample 2D slice. Bottom: MIP.

already visited face-neighbours subtracted by one. Pseudo-code for the forward scan of the $D^6$ reverse DT (the backward scan is similar) [**NB95**]:

```
/*------------------FORWARD SCAN--------------------------------*/
LOOPXYZ(image) {
   if (x<xLO || x>xHI || y<yLO || y>yHI || z<zLO || z>zHI)
     I(x, y, z) = 0; /* image border omitted */
   else
     I(x, y, z) = MAX(I(x,y,z-1)-1,
                      I(x,y-1,z)-1,
                      I(x-1,y,z)-1,
                      I(x, y, z))
}
/*------------------END--FORWARD SCAN--------------------------*/
```

An object voxel is called a *centre of a maximal ball* (CMB), if the associated ball is not completely covered by any other single ball in the object. The union of the maximal balls, which might be computed by applying the reverse DT using the set of CMBs as seed values, coincides with the whole object. For $DT^6$, the CMBs are equivalent to the local maxima. (The situation is more complex for other DTs.)

**3.2. Skeletonization.** The skeletonization method we have used starts with a binary volume image. A number of skeletal properties are fulfilled for our skeletons:

- They have the same topology as the object.
- They preserve the general shape of the object.
- They are centred in the object.
- They are a thin subset of the object.
- They allow reconstruction of the object.

Preserving the topology in this case means that the number of components, the number of tunnels, and the number of cavities all remain the same. In 3D, one must distinguish between two different types of "holes", i.e., tunnels, like what goes through a doughnut, and cavities, like what exists in a hollow ball. Also, in contrast with the 2D case, it is not sufficient to count only the components of the object, but it is necessary to count the components of the background as well. On digital topology, see also [**KR89, MBA93**].

The reconstruction property is true for the surface skeletons, and sometimes also true for the curve skeletons (see below).

The skeletonization method is described in detail in [**BNSdB99**]. It is based on a small number of simple neighbourhood operations; it is fairly time efficient. The skeletonization consists of five parts, where Part I and Part II reduce the original solid 3D object to a surface skeleton consisting of surfaces which are only one voxel thick. In Part III and Part IV, we go further and reduce this surface skeleton to a set of curves which are only one pixel wide. Part V, that may be added, is a "pruning" process, which is performed to remove irrelevant branches.

The method we have used differs in some respects from [**BNSdB99**]. By computing the $DT^6$ of the object first of all, instead of iteratively identifying the border voxels when computing the surface skeleton, we reduce computation time. Border voxels can be directly defined on the $DT^6$ by the distance layers, which equals the current iteration. Each successive border layer is 6-connected to the previous layer, exactly as the layers are ordered in the $DT^6$. Internal voxels are voxels with higher distance values. For details, see [**BNSdB99**]. The time reduction lies in the iterative process, where now only three passes, instead of four, are needed in each iteration:

**Pass 1:** Inspect voxels with the current iteration number. Mark those that are *multiple*, i.e., CMBs and voxels necessary for preserving connectivity. See [**BNSdB99**] for details.

**Pass 2:** Again, inspect voxels with the current iteration number. Mark those that are *tunnel* voxels, which, if removed, would create a tunnel.

**Pass 3:** Remove all unmarked border voxels.

The process terminates when all border voxels are multiple, and hence no more voxels can be removed.

The rules for detecting multiple voxels prevent disconnecting objects of all connectivities, but they do not prevent creation of tunnels in thin complex surfaces. Therefore, Pass 2 is necessary (see [**BNSdB99**]). The original object can be reconstructed from the resulting surface skeleton by growing balls of radius equal to the $DT^6$ values, e.g., by efficiently applying the reverse $DT^6$, as all the CMBs are included. This reconstructibility is lost in the following skeletonization parts.

Part I, which iteratively removes simple (or non-multiple) border voxels, typically results in something which is almost a surface skeleton but not quite, as it may be up to two voxels thick where the original object had a width of an even number of voxels. We therefore have a separate Part II in the algorithm to reduce the thickness from two to one.

Similarly, in the curve skeletonization, Part III removes voxels from the outer edges of the surface until the skeleton is at most two voxels wide:

**Pass 1:** Identify *outer* voxels of the surface, i.e., voxels whose removal would not create tunnels.

**Pass 2:** Mark *protrusion* and *break-point* voxels. See [**BNSdB99**] for details.

**Pass 3:** Sequentially remove all unmarked outer voxels, unless that would disconnect the object.

In case the original surface is an even number of voxels in width, the curve skeleton becomes two-voxel wide. It is reduced to unit thickness by iteratively applying a thinning process in Part IV.

Finally, when the skeleton consists of thin curves, we may post-process the resulting skeleton by removing short branches that are assumed to be caused by noise in the original image. This "pruning" (Part V) is described in Section 4.3. Zig-zags in the curve may also be straightened.

## 4. Resulting curve skeletons

The resulting curve skeleton will always be 26-connected, if the original object was so [**BNSdB99**]. Below, we give two examples: one synthetic image and one MRA image. The figures for our examples are rendered 2D projections of the 3D objects, i.e., segmented data.

**4.1. Synthetic image.** The results of skeletonization of a pyramid (base $85 \times 85$ and height 122 voxels) can be seen in Figure 2. The curve skeleton consists of the meeting points of all surfaces in the surface skeleton, which results in branches starting in each corner going inwards.

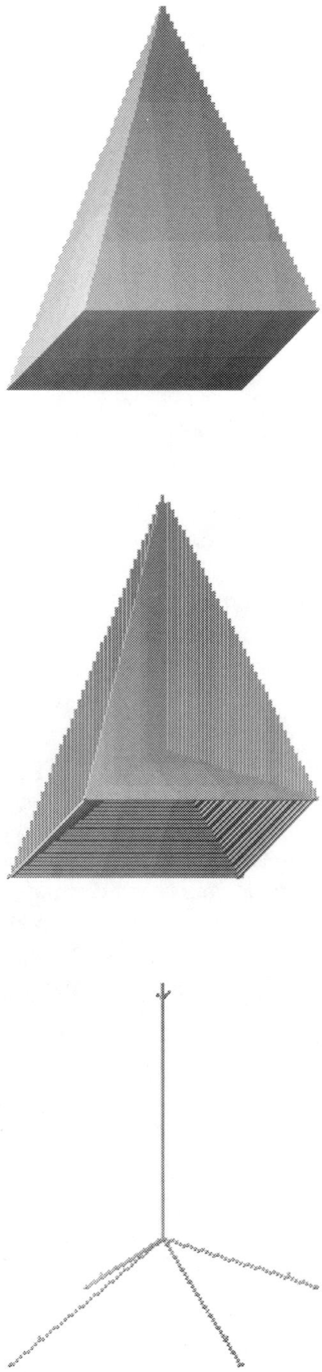

FIGURE 2. A pyramid (top), its surface skeleton (middle), and its curve skeleton (bottom).

**4.2. Magnetic resonance angiography image.** A $512 \times 512 \times 50$ image of the arteries of the pelvis was obtained with Gadolinium MRA technique in a 1.5 T magnet (Philips Gyroscan ACS-NT). The voxel size was $0.759 \times 0.759 \times 1.799$ mm.

The following image processing methods were performed in the analysis of the blood vessels:

1. Interpolating the image to cubic voxels.
2. Segmentation of the blood vessels from the background:
   (a) simple grey-level thresholding
   (b) connected component labelling
   (c) interactive removal of noise
   (d) morphological smoothing operations
   (e) cavity filling
3. Surface skeletonization.
4. Curve skeletonization.
5. Skeleton pruning.
6. Quantitative and qualitative skeleton analysis.

Computations are simpler if distances between layers are the same for all directions $x$, $y$, and $z$. Therefore, the anisotropy of the image is eliminated by performing a linear interpolation of the image to *cuberille* voxels. There exist other (not as simple) solutions to this problem. In this particular case the interpolated image is of size $512 \times 512 \times 117$ (approximately 30 Mbyte). This will make, among other things, the distance calculations and connectivity aspects a lot easier.

In most cases, segmentation is a difficult task. If a correct segmentation of the blood vessels from the background is not obtained, the results may be completely misleading. In our example, though, this problem was fairly easily solved. Due to the Gd-MRA technique, the contrast between the bright arteries and the darker background was high enough to allow a simple grey-level thresholding. If the segmentation would be more difficult the concept in, e.g., [**US96**] could be used.

However, the bright noise in the image is also thresholded, which is undesirable. To eliminate noise, we applied a 3D connected component labelling, in which a volume threshold discards small objects. Hence, the main (largest) component was selected, i.e., the blood vessels. Finally, a morphological closing operation (three iterations of size $3 \times 3 \times 3$) was performed to reduce artificial irregularities in the vessel border.

Next, the skeletonization algorithm was applied. See Figure 3 (top-left). The surface skeleton and the corresponding curve skeleton are shown for the MRA image in Figure 3, at top-right and bottom-left, respectively.

**4.3. Pruning.** When we apply the skeletonization to volume images from the real world and not only to synthetic ones, we soon find that the method is quite sensitive to noise in the input data. Typically, the curve skeleton will be rather "hairy", i.e., it will have a large number of short branches. See Figure 3 (bottom-left). Each of these branches corresponds to "unsmoothness" in the original object, either resulting from significant variations or from noise. We remove these in a *pruning* part (analogous to what is done to an apple tree). However, one must be aware of the fact that this is a rather arbitrary decision. It may be that this irregularity of the skeleton reflects an irregularity of the vessel wall, which might be due to atherosclerosis. In other words, we risk to destroy clinically relevant information

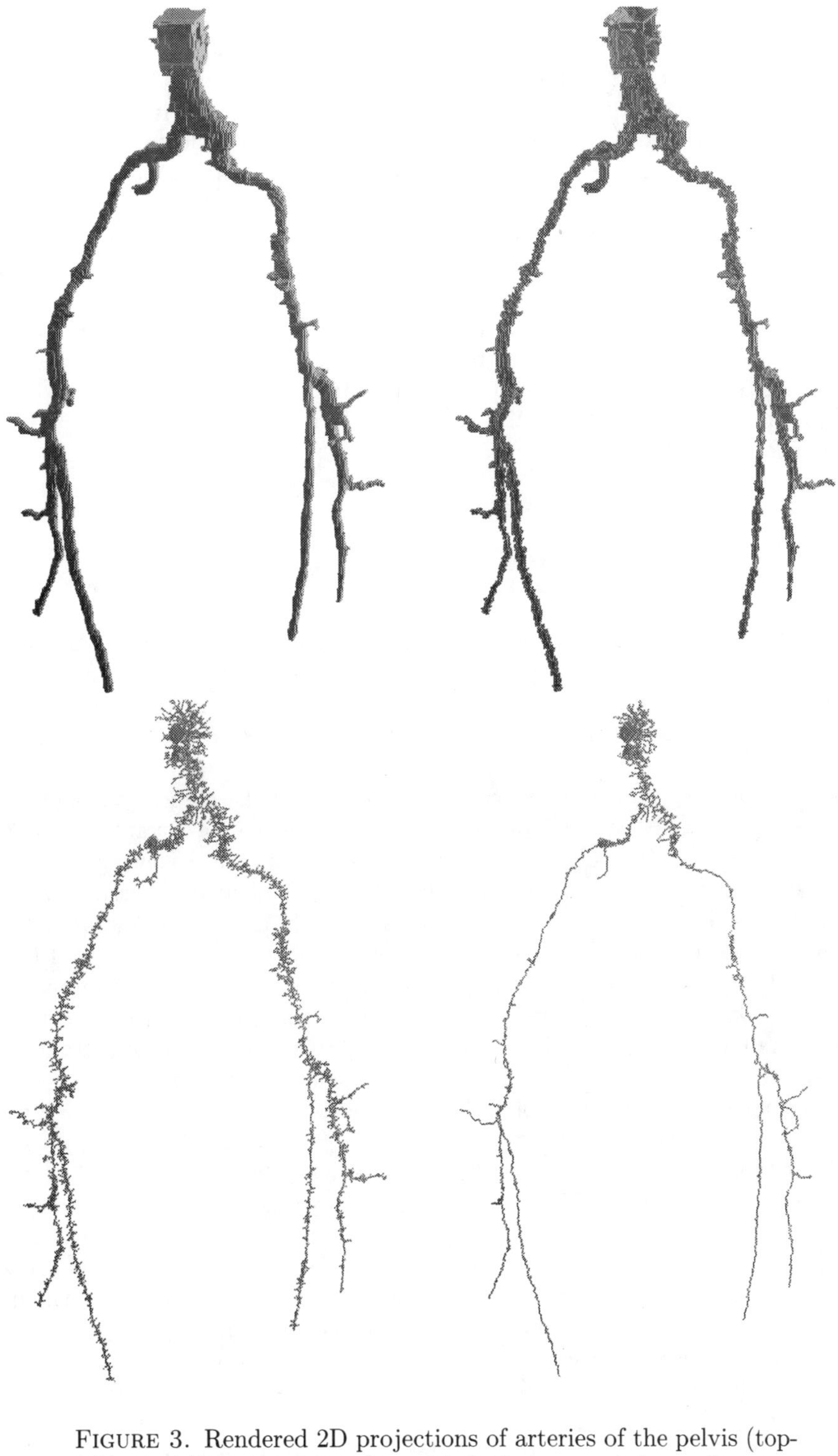

FIGURE 3. Rendered 2D projections of arteries of the pelvis (top-left), the surface skeleton (top-right), the curve skeleton (bottom-left), and the pruned curve skeleton (bottom-right).

by making the skeleton too smooth in the "beautifying" process. Consequently, there is a problem in deciding which branches to remove and which to preserve.

The simple pruning method we have used utilizes the distance values of the curve skeleton. The principle is to cut away all short branches by iteratively removing voxels with distance values less than a certain threshold, unless a removal would disconnect the skeleton. See Figure 3 (bottom-right), where threshold value = 5 has been used for the distance labels. A significant improvement is obtained even with this simple pruning. All branches are shortened by the same number of voxels, so that all the short branches disappear. For the long main branches this little shortening is usually of no importance.

After pruning we have an acceptable representation of the arteries and branches in the pelvis. It may be of interest to study the number of voxels in each object. The number of voxels in the pruned curve skeleton is less than 2.5 % of that of the original object (150 000), but still it gives a fairly good impression of the general shape of the arterial tree.

## 5. Presentation method

A problem with projection images for visualizing 3D angiographic data is that non-circular stenoses are often seen only if the projection is made in a certain direction. For example, a stenosis reducing mainly the anteroposterior diameter of a vessel is difficult to perceive if the projection is made in the frontal plane, simulating anteroposterior rays through the volume. To overcome this limitation, MRA images are often presented as rotation sequences of 2D projections. This approach, however, often implies practical problems in the clinical setting where video display is not always available. One of our goals is to concentrate information about stenoses into one single projection, independent of the viewing direction. Our tool to obtain this is the distance-labelled curve skeleton described above. The skeletal voxels are labelled with their distance to the original background. Each distance value then represents the radius of a ball as large as possible still being a subset of the object. For the surface skeleton, the union of all balls reconstruct the original object. For the curve skeleton, the distance values correspond to the minimum radius of the object at that point, which we will utilize in our presentation method. By surrounding each voxel in the curve skeleton with a ball of radius equal to the distance value, e.g., by applying the reverse DT, the object is displayed with minimum diameter at each point.

The presentation method is first illustrated by a synthetic example. The binary volume image ($256 \times 256 \times 256$ voxels) contains a cylinder of radius 10 voxels and height 240 voxels. Three artificial "stenoses" have been introduced (called upper, middle, and lower). In Figure 4 (left) the object is shown as rendered 2D projections for two different viewing directions. The upper stenosis is visible for the first projection, but not for the second. The middle stenosis should be visible in any projection. The lower stenosis can be seen in the second projection, but is hardly visible in the first (in a MIP it would not be visible). The curve skeleton can be seen in the middle of Figure 4, describing the shape of the original object well. As long as the original surface is a smooth cylinder, the curve skeleton is a simple curve. Where the shape changes skeletal protrusions are introduced. In this application we do not remove any of these branches.

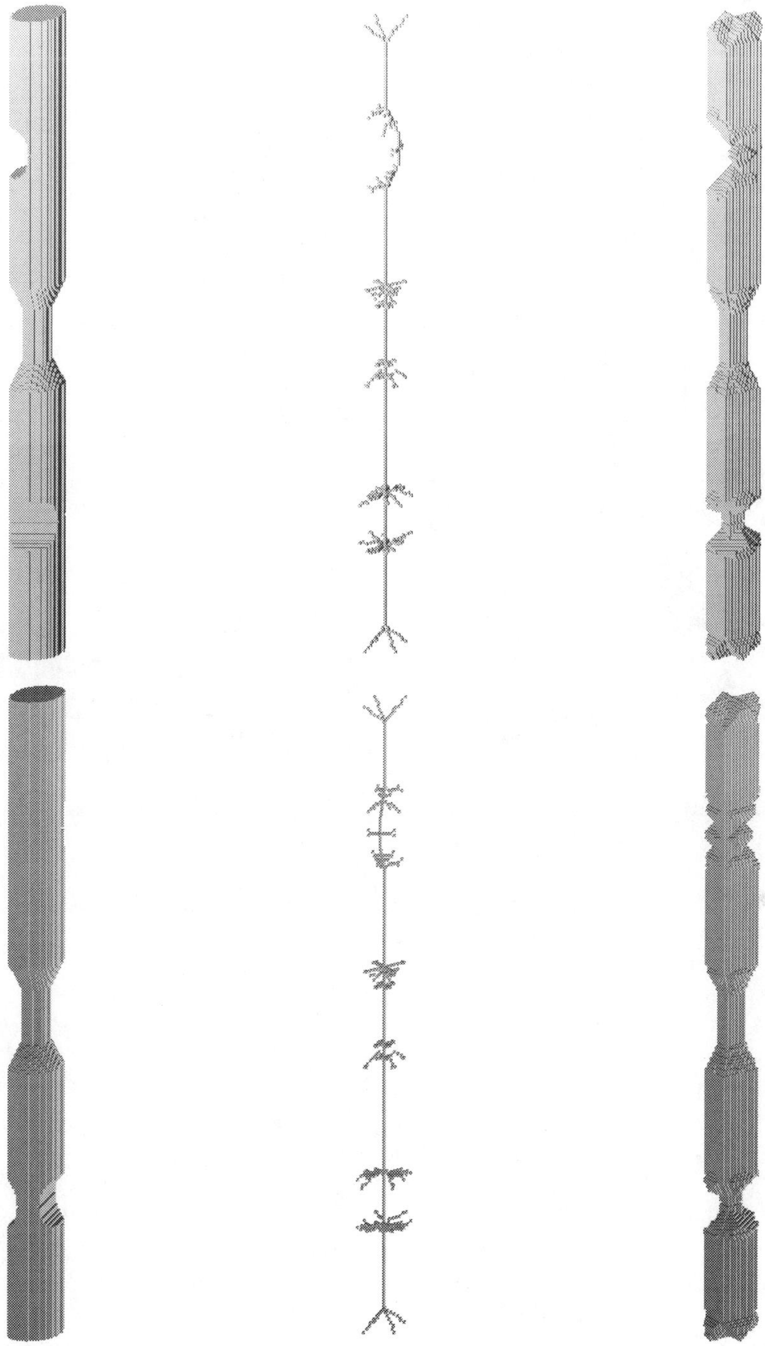

FIGURE 4. A synthetic image with three artificial "stenoses". Left: rendered 2D projections of the object for two different viewing directions. Middle: the corresponding curve skeleton. Right: reconstruction from the curve skeleton, showing that the choice of viewing direction is no longer crucial for detection of the "stenoses."

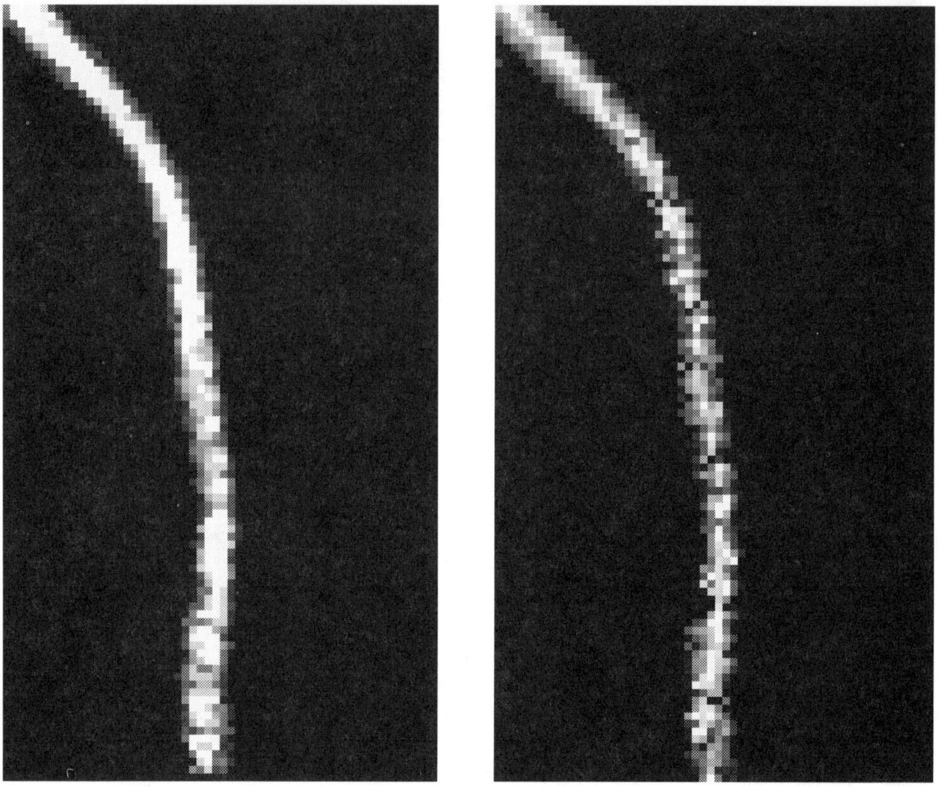

FIGURE 5. Rendered 2D projections of the segmented left external iliac artery before (left) and after (right) curve skeletonization and reverse distance transformation. The lower part of the right image is a possible location for a stenosis, which is not visible in the image to the left.

Finally, the reconstruction is shown to the right in Figure 4. If the reconstruction was performed on the surface skeleton, we would obtain the original object. But as the reverse DT is applied to the curve skeleton, this is not the case. One may notice that the influence of the underlying metric, $D^6$, is evident in the result; a number of octahedra are added together giving the "diamond shape" of the cross-section. More importantly, all three stenoses are now visible in both projections; the reconstruction uses the minimum distance value to the original background at each voxel.

An example of the application of the proposed presentation method on real MRA data is found in Figure 5. Here, the segmented left external iliac artery is shown before and after the process described above. In this case a segment with reduced minimum radius is found in the lower part of the image only in the processed version, indicating that this region should be studied in more detail in the original data set.

## 6. Discussion

It seems that the distance-labelled curve skeleton (after pruning) is a meaningful representation of 3D vessel geometry. In addition to the overall tree anatomy, they contain information about the calibre of the vessels, which is the most interesting feature, at least for the diagnosis of atherosclerosis.

We have also described a method to display "stenoses" in 3D images independent of the viewing direction by combining distance transform based skeletonization with the reverse distance transformation. For example, a stenosis reducing the anteroposterior diameter of a craniocaudal vessel, which is normally seen only in the lateral projection, will then become visible also in the frontal projection. It should be emphasized that these rather diagrammatic sketches are not meant to replace traditional rendering methods, but rather to focus the attention of the viewer on certain regions which should be scrutinized with other methods, e.g., viewing the original slices.

One of the advantages of having a distance transform based skeletonization method, is that reversibility of the object might be possible (depending on the method). Reversibility is important when characterizing objects, otherwise much of the shape information is lost. The reversibility property has hardly been considered in the 3D literature. See [**SNB99**]. Other examples of methods based on iterative thinning in distance transforms are [**ST95**] and [**Pud98**]. However, reversibility is not discussed in these references. Obviously, the dependence of the vessel orientation relative to the coordinate system imposed by the $D^6$ metric is undesirable. An important task for our future research is therefore to develop a skeletonization method based on a less rotation dependent distance transform, more closely approximating Euclidean distance. We are also working to find more intelligent pruning methods which do not shorten main branches.

A serious limitation of this study is that segmented (thresholded) data have been used, since the algorithm has so far been formulated only for binary images. Even though Gd-MRA images are comparatively easy to segment, the difficulty in finding a suitable threshold value must not be underestimated. We plan to combine the presentation approach with the grey-scale connectivity concept developed in a different project for MRA volume visualization [**SSL99**]. Furthermore, any segmented volume image will inevitably contain less information than the grey-scale original. Hence, it is clearly motivated to attempt to transfer the concepts to the grey-scale images, possibly employing some kind of "fuzzy skeleton" built on the concepts from fuzzy digital topology [**Ros79**]. Clearly, this would involve a substantial part of theoretical development.

## Acknowledgment

Algorithmic support on skeletonization and distance transformation was given by Prof. Gunilla Borgefors, Dr. Gabriella Sanniti di Baja, and Stina Svensson, which is gratefully acknowledged.

## References

[AET93] Charles M. Anderson, Robert R. Edelman, and Patrick A. Turski, *Clinical magnetic resonance angiography*, Lippincott-Raven Publishers, New York, 1993.

[Ber95] Gilles Bertrand, *A parallel thinning algorithm for medial surfaces*, Pattern Recognition Letters **16** (1995), 979–986.

[BNSdB99] Gunilla Borgefors, Ingela Nyström, and Gabriella Sanniti di Baja, *Computing skeletons in three dimensions*, Pattern Recognition **32** (1999), no. 7, 1225–1236.

[Bor84] Gunilla Borgefors, *Distance transformations in arbitrary dimensions*, Computer Vision, Graphics, and Image Processing **27** (1984), 321–345.

[Bor96] Gunilla Borgefors, *On digital distance transforms in three dimensions*, Computer Vision and Image Understanding **64** (1996), no. 3, 368–376.

[DHW+97] C. P. Davis, T. F. Hany, S. Wildermuth, et al., *Postprocessing techniques for gadolinium-enhanced three-dimensional MR angiography*, RadioGraphics **17** (1997), no. 5, 1061–1077.

[GSV96] Yaorong Ge, David R. Stelts, and David J. Vining, *3D skeleton for virtual colonoscopy*, Proceedings of $4^{th}$ VBC'96: Visualization in Biomedical Computing (Karl Heinz Höhne and Ron Kikinis, eds.), Springer-Verlag, Berlin Heidelberg, 1996, pp. 449–454.

[KR89] T. Yung Kong and Azriel Rosenfeld, *Digital topology: Introduction and survey*, Computer Vision, Graphics, and Image Processing **48** (1989), 357–393.

[LKC94] Ta-Chih Lee, Rangasami L. Kashyap, and Chong-Nam Chu, *Building skeleton models via 3-D medial surface/axis thinning algorithms*, CVGIP: Graphical Models and Image Processing **56** (1994), no. 6, 462–478.

[MBA93] Grégoire Malandain, Gilles Bertrand, and Nicholas Ayache, *Topological segmentation of discrete surfaces*, International Journal of Computer Vision **10** (1993), no. 2, 183–197.

[MS96] Cherng Min Ma and Milan Sonka, *A fully parallel 3D thinning algorithm and its applications*, Computer Vision and Image Understanding **64** (1996), no. 3, 420–433.

[NB95] Ingela Nyström and Gunilla Borgefors, *Synthesising objects and scenes using the reverse distance transformation in 2D and 3D*, Proceedings of ICIAP'95 - Sanremo: Image Analysis and Processing (Carlo Braccini, Leila DeFloriani, and Gianni Vernazza, eds.), Springer-Verlag, Berlin Heidelberg, 1995, pp. 441–446.

[NMR+92] Sandy Napel, Michael P. Marks, Rubin, et al., *CT angiography with spiral CT and maximum intensity projection*, Radiology **185** (1992), 607–610.

[Nys98] Ingela Nyström, *Skeletonization applied to magnetic resonance angiography images*, Medical Imaging 1998: Image Processing (Kenneth M. Hanson, ed.), Proc. SPIE 3338, 1998, pp. 693–701.

[PK99] Kálmán Palágyi and Attila Kuba, *A parallel 3D 12-subiteration thinning algorithm*, Graphical Models and Image Processing **61** (1999), 199–221.

[Pri94] Martin R. Prince, *Gadolinium-enhanced MR aortography*, Radiology **191** (1994), 155–164.

[Pud98] Chris Pudney, *Distance-ordered homotopic thinning: A skeletonization algorithm for 3D digital images*, Computer Vision and Image Understanding **72** (1998), no. 3, 404–413.

[Ros79] Azriel Rosenfeld, *Fuzzy digital topology*, Information and Control **40** (1979), 76–87.

[RP66] Azriel Rosenfeld and John L. Pfaltz, *Sequential operations in digital picture processing*, Journal of the Association for Computing Machinery **13** (1966), no. 4, 471–494.

[SCM97] Punam Kumar Saha, B. B. Chaudhuri, and Dwijesh Dutta Majumder, *A new shape preserving parallel thinning algorithm for 3D digital images*, Pattern Recognition **30** (1997), no. 12, 1939–1955.

[SNB99] Stina Svensson, Ingela Nyström, and Gunilla Borgefors, *Fully reversible skeletonization for volume images based on anchor-points from the $D^{26}$ distance transform*, Proceedings of $11^{th}$ Scandinavian Conference on Image Analysis, Kangerlussuaq, Greenland, June 1999, pp. 601–608.

[SSL99] Ö Smedby, Stina Svensson, and Tomas Löfstrand, *Greyscale connectivity concept for visualizing MRA and CTA volumes*, Medical Imaging 1999: Image Display (Seong Ki Mun and Yongmin Kim, eds.), Proc. SPIE 3658, 1999, pp. 212–219.

[ST95] Toyofumi Saito and Jun-Ichiro Toriwaki, *A sequential thinning algorithm for three dimensional digital pictures using the Euclidean distance transformation*, Proceedings of $9^{th}$ Scandinavian Conference on Image Analysis, Uppsala, Sweden, 1995, pp. 507–516.

[US96] Jayaram K. Udupa and Supun Samarasekera, *Fuzzy connectedness and object definition: Theory, algorithms, and applications in image segmentation*, Graphical Models and Image Processing **58** (1996), no. 3, 246–261.

Centre for Image Analysis, Uppsala University, Uppsala, Sweden
*E-mail address*: Ingela.Nystrom@cb.uu.se
*URL*: http://www.cb.uu.se/~ingela

Division of Medical Radiology, University Hospital, Linköping, Sweden
*E-mail address*: Orjan.Smedby@imv.liu.se

DIMACS Series in Discrete Mathematics
and Theoretical Computer Science
Volume **55**, 2000

# Computing Medial Surfaces

Sylvain Bouix and Kaleem Siddiqi

ABSTRACT. The medial surface of a volumetric object is of significant interest for shape analysis in bio-medicine. However, its numerical computation can be subtle. Methods based on Voronoi techniques provide connected structures which preserve the object's topology, but heuristic pruning measures are introduced to remove unwanted faces. Approaches based on Euclidean distance functions can localize medial surface points accurately, but often at the cost of altering the object's topology. In this paper we present an overview of a new algorithm for computing medial surfaces which addresses these concerns. The method is robust and accurate, has low computational complexity, and preserves topology. The key idea is to measure the net outward flux of a vector field per unit volume, and to detect locations where a conservation of energy principle is violated. This is done in conjunction with a thinning process applied on a cubic lattice. We illustrate the algorithm with examples of medial surfaces of synthetic objects as well as complex anatomical structures obtained from medical images.

## 1. Introduction

Medial surface based representations are of significant interest for a number of applications in biomedicine, including object representation [**21, 31**], registration [**17**] and segmentation [**27**]. Such descriptions are also popular for animating objects in graphics [**32, 24**] and manipulating them in computer-aided design. They provide a compact representation while preserving the object's genus and retain sufficient local information to reconstruct (a close approximation to) it. This facilitates a number of important tasks including the quantification of the local width of a complex structure, e.g., that of the grey matter in the human brain, and the analysis of its topology, e.g., that of blood vessels in angiography images. Graph-based abstractions of such data have also been proposed [**11**]. Despite their popularity, the stable numerical computation of medial surfaces remains a challenging problem. Unfortunately, the classical difficulties associated with computing their 2D analog, the Blum skeleton, are only exacerbated when a third dimension is added.

2000 *Mathematics Subject Classification.* 92C55.
*Key words and phrases.* 3D Shape Analysis, Medial Surfaces, Skeletons, Medical Image Understanding.

**1.1. Background.** The 2D skeleton of a closed set $A \subset \mathcal{R}^2$ is the locus of centers of maximal open discs contained within the complement of the set [**6**]. An open disc is maximal if there exists no other open disc contained in the complement of $A$ that properly contains the disc. The medial surface of a closed set $A \subset \mathcal{R}^3$ is defined in an analogous fashion as the locus of centers of maximal open spheres contained in the complement of the set. It is often referred to as the 3D skeleton, though this term is misleading since it is in fact comprised of a collection of 3D points, curves and surfaces [**3**]. Whereas the above definition is quite general, in the current context we shall assume that the closed set $A$ is the bounding surface of a volumetric object. Hence, this set will have two complementary medial surfaces, one inside the volume and the other outside it. In most cases we shall be referring to the former, though the development applies to both.

Interest in the medial surface as a representation for a volumetric object stems from a number of interesting properties: i) it is a *thin* set, i.e., it contains no interior points, ii) it is homotopic to the volume, iii) it is invariant under Euclidean transformations of the volume (rotations and translations), and iv) given the radius of the maximal inscribed sphere associated which each medial surface point, the volumetric object can be reconstructed exactly. Hence, it provides a compact representation while preserving the object's genus and making certain useful properties explicit, such as its local width.

Approaches to computing skeletons and medial surfaces can be broadly organized into three classes. First, methods based on *thinning* attempt to realize Blum's grassfire formulation [**6**] by peeling away layers from an object, while retaining special points [**2, 15, 20**]. It is possible to define erosion rules on a lattice such that the topology of the object is preserved. However, these methods are quite sensitive to Euclidean transformations of the data and typically fail to localize skeletal or medial surface points accurately. As a consequence, only a coarse approximation to the object is usually reconstructed [**20, 5, 15**].

Second, it has been shown that under appropriate smoothness conditions, the vertices of the Voronoi diagram of a set of boundary points converges to the exact skeleton as the sampling rate increases [**26**]. This property has been exploited to develop skeletonization algorithms in 2D [**22**], as well as extensions to 3D [**28, 29**]. The dual of the Voronoi diagram, the Delaunay triangulation (or tetrahedralization in 3D) has also been used extensively. Here the skeleton is defined as the locus of centers of the circumsribed spheres of each tetrahedra [**13, 21, 4**]. Both types of methods preserve topology and accurately localize skeletal or medial surface points, provided that the boundary is sampled densely. Unfortunately, however, the techniques used to prune faces and edges which correspond to small perturbations of the boundary are typically based on heuristics. In practice, the results are not invariant under Euclidean transformations and the optimization step, particularly in 3D, can have a high computational complexity [**21**].

A third class of methods exploits the fact that the locus of skeletal or medial surface points coincides with the singularities of a *Euclidean distance function* to the boundary. These approaches attempt to detect local maxima of the distance function, or the corresponding discontinuities in its derivatives [**1, 16, 25, 14**]. The numerical detection of these singularities is itself a non-trivial problem; whereas it may be possible to localize them, ensuring homotopy with the original object is difficult.

In this paper we present an overview of a novel algorithm for computing medial surfaces which addresses several of the above concerns. The method is robust and accurate, has low computational complexity, and preserves topology. The key idea is to measure the net outward flux per unit volume of the gradient vector field of the Euclidean distance function, and to detect locations where a conservation of energy principle is violated. This is motivated by the observation that the grassfire flow leads to a hamilton-jacobi equation, which by nature is conservative in the smooth regime of its underlying phase space. This formulation was introduced in our earlier work, in the context of simulating the eikonal equation [**30**]. We have combined a measurement of the flux with a topology preserving thinning process applied on a cubic lattice, such that the result is homotopic to the original object [**10**]. Our goal in this paper is to introduce these ideas to a different community. There are other recent methods which combine aspects of thinning, Voronoi diagrams and distance functions [**19, 34, 9, 33**]. In spirit, our method is closest to that of [**19**] [1] but is grounded in principles from physics. We illustrate the algorithm with several examples including medial surfaces of complex anatomical structures obtained from medical images.

## 2. Hamiltonian Medial Surfaces

Consider the grassfire flow

$$\frac{\partial S}{\partial t} = \mathcal{N} \tag{2.1}$$

acting on a 3D surface $S$, such that each point on its boundary is moving with unit speed in the direction of the inward normal $\mathcal{N}$. In recent work, we have shown that this formulation leads to a Hamilton-Jacobi equation on the Euclidean distance function to the initial surface [**30**]. In physics, such equations are typically solved by looking at the evolution of the phase space of the equivalent Hamiltonian system. Since Hamiltonian systems are conservative, the locus of medial surface points coincides with locations where a conservation of energy principle is violated. This loss of energy can be used to formulate a natural criterion for detecting singularities of the distance function.

In more formal terms, let $D$ be the Euclidean distance function to the initial surface $\mathcal{S}_0$. The magnitude of its gradient, $\|\nabla D\|$, is identical to 1 in its smooth regime. With $\mathbf{q} = (x, y, z)$, $\mathbf{p} = (D_x, D_y, D_z)$, the Hamiltonian system is given by

$$\dot{\mathbf{p}} = -\frac{\partial H}{\partial \mathbf{q}} = (0,0,0), \qquad \dot{\mathbf{q}} = \frac{\partial H}{\partial \mathbf{p}} = -(D_x, D_y, D_z), \tag{2.2}$$

with an associated Hamiltonian function $H = 1 - (D_x^2 + D_y^2 + D_z^2)^{\frac{1}{2}}$. It is straightforward to show that all Hamiltonian systems are conservative [**23**, p. 172]; hence they are divergence free. Conversely, when trajectories of the system intersect, a conservation of energy principle is violated. This suggests a natural approach for locating the medial surface: compute the divergence of the gradient vector field $\dot{\mathbf{q}}$ and detect locations where it is not zero. The divergence is defined as the net

[1] Malandain and Fernandez-Vidal use a heuristic estimation of the singularities of a distance function to obtain an initial skeleton or medial surface, and then perform a topological reconstruction to ensure homotopy with the original shape.

outward flux per unit volume, as the volume about the point shrinks to zero:

$$\text{div}(\dot{\mathbf{q}}) \equiv \lim_{\Delta \text{v} \to 0} \frac{\int_{\text{S}} < \dot{\mathbf{q}}, \mathcal{N} > \text{ds}}{\Delta \text{v}} \tag{2.3}$$

Here $\Delta v$ is the volume, $S$ is its bounding surface and $\mathcal{N}$ is the outward normal at each point on $S$. Via the divergence theorem

$$\int_{v} \text{div}(\dot{\mathbf{q}})\text{dv} \equiv \int_{\text{S}} < \dot{\mathbf{q}}, \mathcal{N} > \text{ds}. \tag{2.4}$$

In other words, the integral of the divergence of the vector field within a finite volume gives the net outward flux through the surface which bounds it. Locations where the flux is negative, and hence energy is lost, correspond to sinks or medial surface points of the interior. Similarly, locations where the flux is positive correspond to sources or medial surface points of the exterior.[2] The above analysis applies in 2D as well: simply replace the initial closed surface $\mathcal{S}$ with a closed contour $\mathcal{C}$ in Eq. (2.1), remove the third coordinate $z$ of the phase space in Eq. (2.2), and replace the volume element with an area element and the surface integral with a contour integral in Eq. (2.3).

We now have a robust method for localizing medial surface points by discretizing Eq. 2.4, and thresholding to select points with negative total outward flux. However, since the computation is local, global properties are not necessarily preserved. In earlier work, we have observed that the method gives accurate medial surfaces, but that as the threshold is varied new holes or cavities may be introduced and the medial surface may get disconnected. The remedy, as we shall now show, is to introduce additional constraints along the lines of those incorporated in [**19**] to ensure that the topology of the object is preserved. The main idea is to incorporate a homotopy preserving thinning process, where the removal of points is guided by their divergence values.

## 3. Homotopy Preserving Medial Surfaces

Our goal is to combine the divergence computation with a thinning process acting on a cubic lattice, such that as many points as possible are removed without altering the object's topology. A point is called a *simple* point if its removal does not change the topology of the object. Hence in 3D, its removal must not disconnect the object, create a hole, or create a cavity. We shall adopt a formal definition of a simple point introduced by Malandain et al. [**18**]. First we review a few basic concepts in digital topology.

**3.1. Digital Topology.** In 3D digital topology, the input is a binary (foreground and background) image stored in a 3D array. We shall consider only cubic lattices, where a point is viewed as a unit cube with 6 faces, 12 edges and 8 vertices. For each point, three types of neighbours are defined:

- *6-neighbours*: two points are 6-neighbours if they share a face,
- *18-neighbours*: two points are 18-neighbours if they share a face or an edge, and

[2]Note that the classical definition of the divergence as the sum of the partial derivatives of the vector field with respect to its component directions cannot be used where it is singular, and these are precisely the points we are interested in.

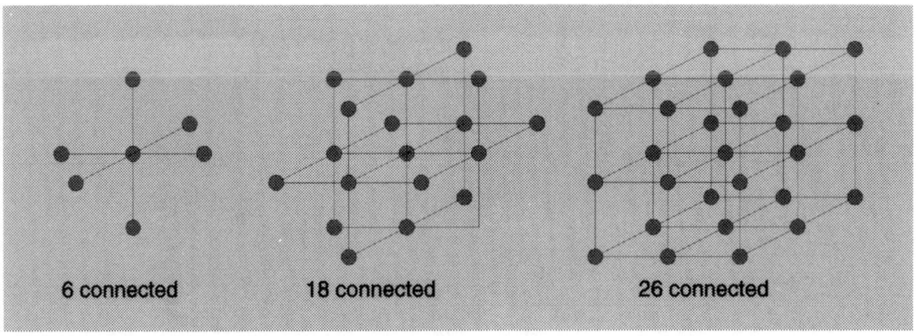


FIGURE 1. 6-neighbourhoods, 18-neighbourhoods and 26-neighbourhoods in a cubic lattice.

- *26-neighbours*: two points are 26-neighbours if they share a face, an edge or a vertex.

The above definitions induce three types of connectivity, denoted *n-connectivity*, where $n \in \{6, 18, 26\}$, as well as three different *n-neighbourhoods* for $x$, called $N_n(x)$ (see Figure 1). A n-neighborhood without its central point is defined as $N_n^* = N_n(x) \backslash \{x\}$. A few more definitions are needed to characterize simple points:

- An object A is *n-adjacent* to an object B, if there exist two points $x \in A$ and $y \in B$ such that $x$ is an n-neighbour of $y$.
- A *n-path* from $x_1$ to $x_k$ is a sequence of points $x_1, x_2, ..., x_k$, such that for all $x_i$, $1 < i \leq k$, $x_{i-1}$ is n-adjacent to $x_i$.
- An object represented by a set of points $O$ is *n-connected*, if for every pair of points $(x_i, x_j) \in O \times O$, there is a n-path from $x_i$ to $x_j$.

Based on these definitions, Malandain et al. provide a topological classification of a point $x$ in a cubic lattice by computing two numbers [**18**]: i) $C^*$: the number of 26-connected components 26-adjacent to $x$ in $O \cap N_{26}^*$, and ii) $\bar{C}$: the number of 6-connected components 6-adjacent to $x$ in $\bar{O} \cap N_{18}$. An important result with respect to our goal of thinning is that if $C^* = 1$ and $\bar{C} = 1$, the point is *simple*, and hence removable. When ensuring homotopy is the only concern, simple points can be removed sequentially until no more simple points are left. The resulting set will be thin and homotopic to the shape. However, the relationship to the medial surface will be uncertain since the locus of surviving points will depend entirely on the order in which the simple points have been removed. In the current context, we have derived a natural criterion for ordering the thinning, based on the divergence of the gradient vector field of the Euclidean distance function.

**3.2. Divergence-Ordered Thinning.** Recall from Section 2, that a conservation of energy principle is violated at medial surface points. The total outward flux of the gradient vector field of the Euclidean distance function is negative at such points, since they correspond to sinks.[3] More importantly, the magnitude of the total outward flux is proportional to the amount of energy absorbed, and hence provides a natural measure of the "strength" of a medial surface point, which we shall use for numerical computations. The essential idea is to order the thinning

[3]Conversely, medial surface points of the background correspond to sources, with positive total outward flux.

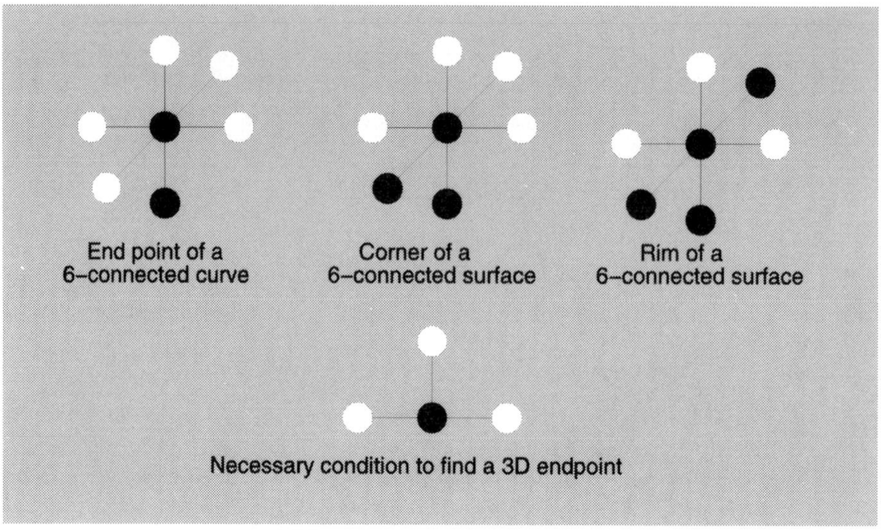


FIGURE 2. An endpoint is defined as the end of a 6-connected curve or the corner or rim of a 6-connected surface in 3D. For each configuration, there exists at least one plane in which the point has at least three background 6-neighbours.

such that the weakest points are removed first, and to stop the process when all surviving points are not simple, or have a total outward flux below some chosen (negative) value, or both. This will accurately localize the medial surface, and also ensure homotopy with the original object. Unfortunately the result is not guaranteed to be a thin set, i.e., one without an interior.

One way of satisfying this last constraint is to define an appropriate notion of an endpoint on a cubic lattice. Such a point would correspond to the endpoint of a curve, or a point on the rim of a surface, in 3D. The thinning process would proceed as before, but the threshold criterion for removal would be applied *only to endpoints*. Hence, all surviving points which are not endpoints would not be simple, and the result would be a thin set.

To facilitate this task, we shall restrict our definition of an endpoint to a 6-connected neighbourhood. In other words, an endpoint is either the end of a 6-connected curve, or a corner or point on the rim of a 6-connected surface. It is straightforward to enumerate the possible 6-connected neighborhoods of the endpoint, and to show that they fall into one of three configurations (see Figure 2). Notice that for each configuration, there exists at least one plane in which the point has at least three background 6-neighbours. This gives us a necessary condition to determine if a point is an endpoint according to our definition. Note that before performing this check, one must also verify that the point is simple.

**3.3. The Algorithm.** The thinning process can be made very efficient by observing that a point which does not have at least one background point as an immediate neighbor cannot be removed, since this would change the object's topology by creating a cavity. Therefore, the only potentially removable points are on the border of the object. Once a border point is removed, only its neighbors are possible candidates for removal. This suggests an implementation using a heap. The complete algorithm can now be stated as follows:

**1** Compute the divergence map.
  **1.1** Compute $D$, the signed distance function to the object [**7**].
  **1.2** Compute $\nabla D$, the gradient vector field of the distance function.
  **1.3** Compute the net outward flux of $\nabla D$ using a discretization of Eq. 2.4.
  For each point $P$, $Flux(P) = \sum_{i=1}^{26} < N_i, \nabla D(P_i) >$,
  where $P_i$ is a 26-neighbour of $P$ and $N_i$ is the
  outward normal at $P_i$ of the unit sphere centered at $P$.

**2** REPEAT
  **2.1** Get the next best candidate point $P$ for removal
  (the point with maximum divergence value on the boundary of the object).
  Note that only points on the boundary need to be considered, since the
  removal of an interior point will create a cavity.
  **2.2** IF $(Flux(P) > Threshold)$ AND ($P$ is *simple*)
  Remove point.
  Go to 2.1
  **2.3** IF $(Flux(P) \leq Threshold)$ AND ($P$ is *simple*) AND ($P$ is not an endpoint)
  Remove point.
  Go to 2.1
  **2.4** ELSE
  Go to 2.1
  UNTIL no more points can be removed.

We can now analyze the complexity of the algorithm. The computation of the distance transform [**7**], the gradient vector field and the total outward flux are all $\mathcal{O}(n)$ operations. Here $n$ is the total number of points in the 3D array. The implementation of the thinning is more subtle. We claim a $\mathcal{O}(klog(k))$ worst case complexity, where $k$ is the number of points in the volume. The explanation is as follows. At first only the points that are on the outer layer of the object are stored in a heap, using the total outward flux as the sorting key for insertion. The extraction of the maximum from the heap will provide the best candidate for removal. If this point is removable it is deleted from the object and its *simple* (potentially removable) neighbours are added to the heap. A point can only be inserted a constant number of times (at most 26 times for a 26-neighbourhood), and insertion in a heap, as well as the extraction of the minimum, are both $\mathcal{O}(log(l))$ operations, where $l$ is the number of elements in the heap. There cannot be more than $k$ elements in the heap, because we only have a total of $k$ points in the volume. The worst case complexity for thinning is therefore $\mathcal{O}(klog(k))$. Hence, the complexity of the algorithm is $\mathcal{O}(n) + \mathcal{O}(klog(k))$.

## 4. Examples

We illustrate the algorithm with both synthetic data and volumes segmented from MR and MRA images. The only free parameter is the choice of the divergence threshold below which the removal of endpoints is blocked. For all examples, this was selected so that approximately 25% of the points within the volume had a lower divergence value.

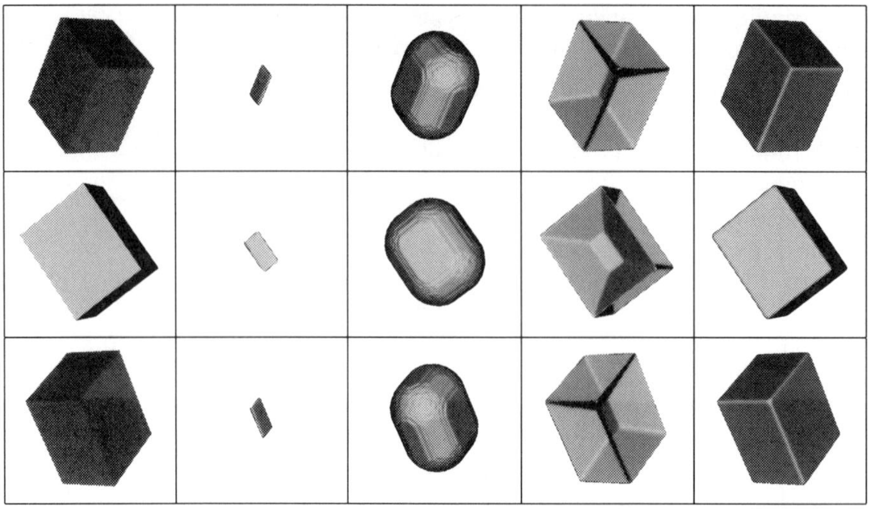

FIGURE 3. FIRST COLUMN: Three views of a cube. SECOND COLUMN: The corresponding medial surfaces computed using the algorithm of [**20**]. THIRD COLUMN: The object reconstructed from the medial surfaces in the previous column. FOURTH COLUMN: The corresponding divergence-based medial surfaces. FIFTH COLUMN: The object reconstructed from the medial surfaces in the previous column.

Figures 3 and 4 compare our approach with the parallel thinning method introduced by Manzanera et al. [**20**]. The results reveal that both frameworks are robust, and yield structures that are homotopic to the underlying object. However, note that the latter method yields only a subset of the "true" medial surface for these data sets, and hence only a coarse approximation to the object is possible. In contrast, a near perfect reconstruction is possible from the divergence-based medial surfaces.

Recall from Section 3 that the computation of the distance transform, its gradient vector field and the total outward flux are all $\mathcal{O}(n)$ operations. Hence, one may be tempted to simply threshold the divergence map below a certain (negative) value to obtain a medial surface extremely efficiently. Unfortunately, due to discretization, the choice of an appropriate threshold parameter is elusive. On the one hand, if the threshold is too high, slight perturbations of the boundary will be represented, and the resulting structure will not be a thin set. On the other hand, lowering the threshold can provide a thin set, but at the cost of altering the object's topology. This is illustrated in Figure 5 (second column) where the medial surfaces corresponding to the views in the first column are accurate and thin, but have holes. This motivates the need for the topological constraints along with the characterization of endpoints discussed in Section 3. Observe that with the same threshold as before, the divergence-based thinning algorithm now yields a thin structure which preserves topology, Figure 5 (third column). The ventricles reconstructed from the medial surfaces in the fourth column are shown in the fifth column.

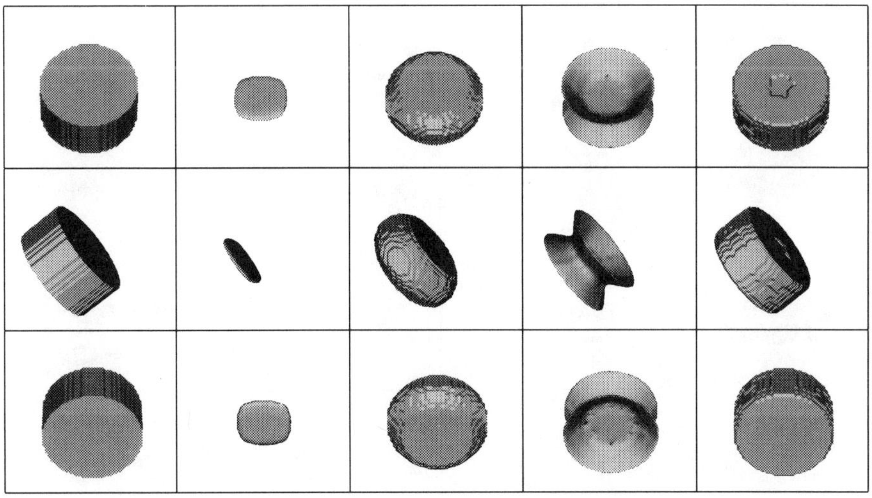

FIGURE 4. FIRST COLUMN: Three views of a cylinder. SECOND COLUMN: The corresponding medial surfaces computed using the algorithm of [**20**]. THIRD COLUMN: The object reconstructed from the medial surfaces in the previous column. FOURTH COLUMN: The corresponding divergence-based medial surfaces. FIFTH COLUMN: The object reconstructed from the medial surfaces in the previous column.

Next, we illustrate the robustness of the approach on a data set of blood vessels obtained from an MRA image of the brain, in Figure 6. The blood vessels have complex topology with loops (due to pathologies), and are already quite thin in several places. The bottom row illustrates the accuracy of the method, where the medial surfaces are shown embedded within the original data set. Generically these structures are thin sheets which approach 3D curves when the blood vessels become perfectly cylindrical. In a number of medical applications where the objects are tubular structures, an explicit reduction of the medial surface to a set of 3D curves is of interest [**12, 8, 9, 34**]. There is a straightforward modification of our framework which allows this, provided that certain special points on the medial surface have been identified. The essential idea is to preserve such special points, but to remove all other simple points in a sequence ordered by their divergence values. This is illustrated for a portion of the vessel data in Figure 7, where the endpoints of 3 branches were selected as special points. Observe that the result is now composed of three 1 voxel wide 26-connected 3D digital curves.

As a final example, Figure 8 illustrates the medial surface of the sulcii of a brain, where we have shown an X, Y and Z slice through the volume. Observe that the medial surface is well localized, and captures the complex topology of the object's shape.

## 5. Conclusions

We have presented an overview of a novel algorithm for computing medial surfaces which is robust and accurate, computationally efficient, invariant to Euclidean

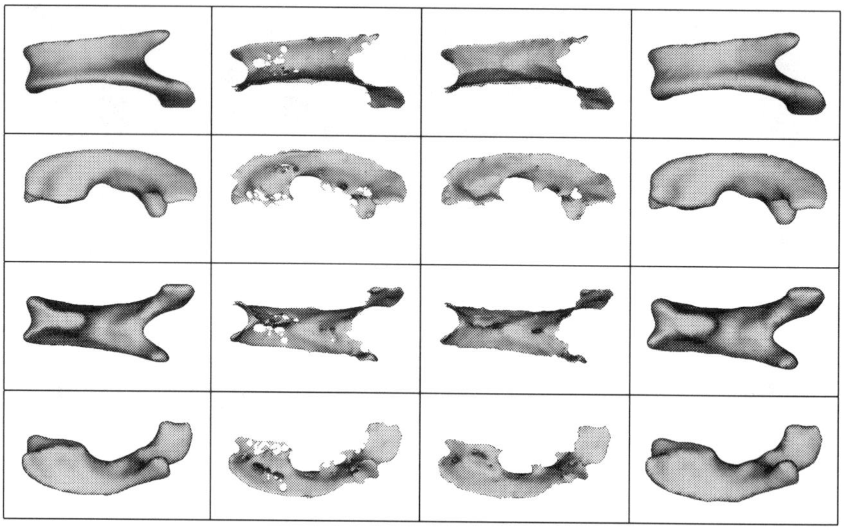

FIGURE 5. FIRST COLUMN: Three views of the ventricles of a brain, segmented from volumetric MR data using an active surface. SECOND COLUMN: The corresponding medial surfaces obtained by thresholding the divergence map. THIRD COLUMN: The divergence-based medial surfaces obtained using the same threshold, but with the incorporation of homotopy preserving thinning. FOURTH COLUMN: The ventricles reconstructed from the divergence-based medial surfaces in the previous column.

transformations and homotopy preserving. The essential idea is to combine a divergence computation on the gradient vector field of the Euclidean distance function to the object's boundary with a thinning process that preserves topology. The characterization of simple (or removable) points is adopted from [**18**], but we have also introduced the notion of an endpoint of a 6-connected structure, in order that the algorithm may converge to a thin set. We have illustrated the advantages of the approach using both synthetic and medical data sets of varying complexity.

We note that in related work, Malandain and Fernandez-Vidal obtain two sets based on thresholding a function of two measures, $\phi$ and $d$, to characterize the singularities of the Euclidean distance function [**19**]. The first set preserves topology but captures many unwanted details and is not thin, while the second set provides a better approximation to the skeleton or medial surface, but often alters the object's topology. The two sets are combined using a topological reconstruction process. Whereas empirical results have been good, the choice of appropriate thresholds are context dependent. More seriously, $\phi$ and $d$, as well as their combination, are all based on heuristics.

In contrast, our method is rooted in a physics-based analysis of the gradient vector field of the Euclidean distance function, which shows that a conservation of energy principle is violated at medial surface points. This justifies the use of the divergence theorem to compute the total outward flux of the vector field, and to locate points where energy is absorbed. It should be clear that whereas we have focussed on the interior of an object, the medial surface of the background can be

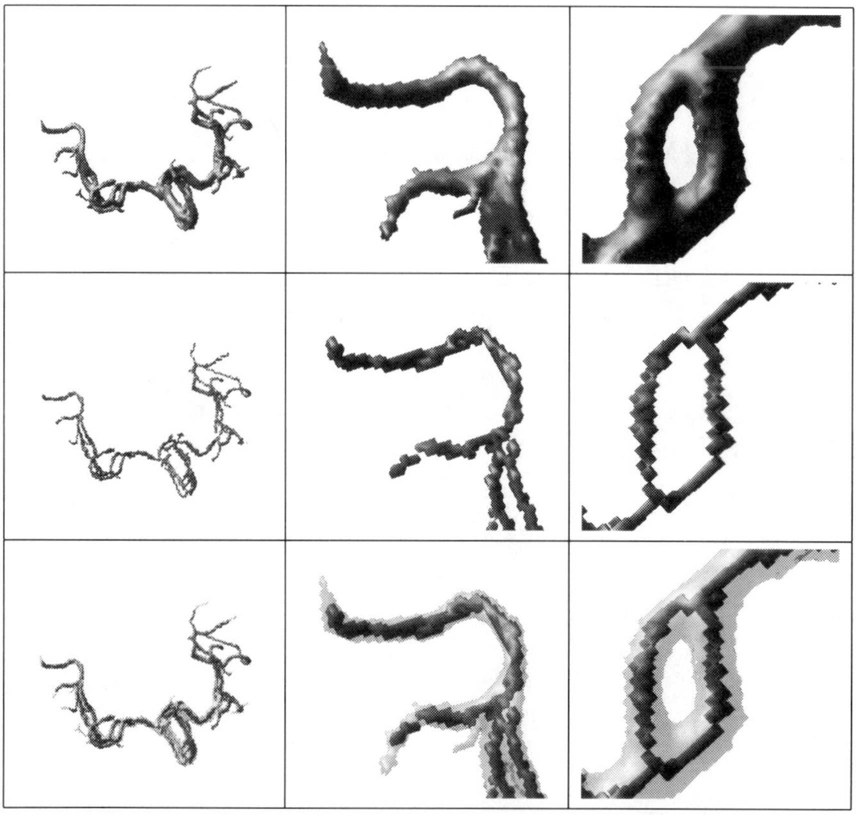

FIGURE 6. TOP ROW: Blood vessels segmented from volumetric MRA data, with magnified parts shown in the middle and right columns. MIDDLE ROW: The corresponding divergence-based medial surfaces. THIRD ROW: The divergence-based medial surfaces (solid) are shown within the vessel surfaces (transparent).

similarly obtained by locating points that act as sources, and have positive total outward flux. Furthermore, both medial surfaces can be located with sub-pixel accuracy by using the local gradient vector field to displace the final set of digital points. In related work we have demonstrated this idea for 2D shapes, where a very similar framework was used to compute 2D skeletons.

In future work we plan to incorporate the topological classification of [**18**] to parse the medial surface and obtain a more abstract representation of it, e.g., as a graph. We shall also explore the possibility of finding necessary as well as sufficient conditions for defining endpoints on a cubic lattice.

**Acknowledgements.** We are grateful to Allen Tannenbaum and Steve Zucker for collaborations on the hamilton-jacobi formulation. Louis Collins, Georges Le Goualher, Belinda Lee and Terry Peters kindly supplied the medical data. This research was supported by CFI, FCAR and NSERC.

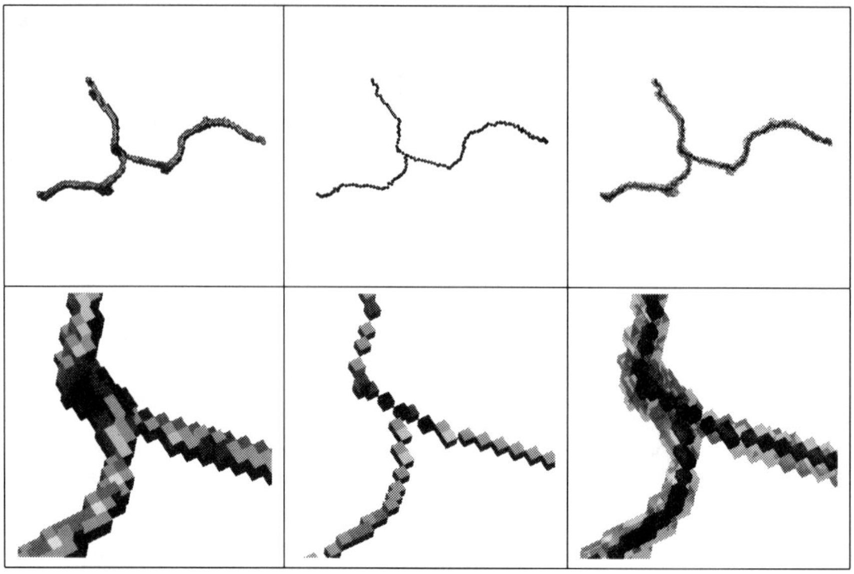

FIGURE 7. LEFT COLUMN: Blood vessels segmented from volumetric MRA data, with a magnified portion shown in the second row. MIDDLE COLUMN: The divergence-based 3D curves. RIGHT COLUMN: The divergence-based 3D curves are shown embedded within the vessel data.

## References


1. C. Arcelli and Gabriela Sanniti di Baja, *Ridge points in euclidean distance maps*, Pattern Recognition Letters **13** (1992), no. 4, 237–243.
2. C. Arcelli and G. Sanniti di Baja, *A width-independent fast thinning algorithm*, IEEE PAMI **7** (1985), no. 4, 463–474.
3. V. I. Arnold, *Mathematical methods of classical mechanics*, Springer-Verlag, 1989.
4. Dominique Attali and Annick Montanvert, *Computing and simplifying 2d and 3d continuous skeletons*, Computer Vision and Image Understanding **67** (1997), no. 3, 261–273.
5. Gilles Bertrand, *A parallel thinning algorithm for medial surfaces*, Pattern Recognition Letters **16** (1995), 979–986.
6. Harry Blum, *Biological shape and visual science*, Journal of Theoretical Biology **38** (1973), 205–287.
7. Gunilla Borgefors, *Distance transformations in arbitrary dimensions*, CVGIP **27** (1984), 321–345.
8. Gunilla Borgefors, Ingela Nystrom, and Gabriella Sanniti Di Baja, *Skeletonizing volume objects part ii: From surface to curve skeleton*, SSPR'98, 1998, pp. 220–229.
9. ———, *Computing skeletons in three dimensions*, Pattern Recognition **32** (1999), 1225–1236.
10. Sylvain Bouix and Kaleem Siddiqi, *Divergence-based medial surfaces*, ECCV'2000 (Dublin, Ireland), June 2000.
11. Elizabeth Bullitt, Stephen Aylward, Alan Liu, Jeffrey Stone, Suresh K. Mukherjee, Chris Coffey, Guido Gerig, and Stephen M. Pizer, *3d graph description of the intracerebral vasculature from segmented mra and tests of accuracy by comparison with x-ray angiograms*, IPMI'99, 1999, pp. 308–321.
12. Y. Ge, D. R. Stelts, X. Zha, D. Vining, and J. Wang, *Computing the central path of the colon from ct images*, SPIE International Symposium on Medical Imaging (San Francisco), vol. 3338(1), 1998, pp. 702–713.

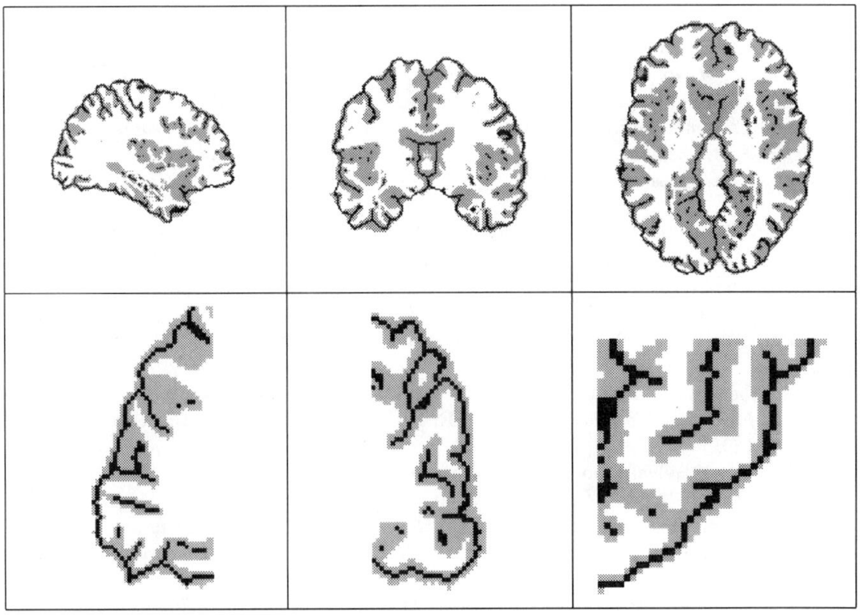

FIGURE 8. TOP ROW: Medial surfaces of the sulcii of a brain, segmented from an MR image. The three columns represent X, Y and Z slices through the volume. The cross section through the medial surface in each slice is shown in black, and the object is shown in grey. BOTTOM ROW: A zoom-in on a selected region of the corresponding slice in the top row, to show detail.


13. John A. Goldak, Xinhua Yu, and Alan Knight abd Lingxian Dong, *Constructing discrete medial axis of 3-d objects*, Int. Journal of Computational Geometry and Applications **1** (1991), no. 3, 327–339.
14. Jose Gomez and Olivier Faugeras, *Reconciling distance fucntions and level sets*, Tech. Report TR3666, INRIA, April 1999.
15. Ta-Chi Lee and Rangasami L. Kashyap, *Building skeleton models via 3-d medial surface/axis thinning algorithm*, CVGIP: Graphical Models and Image Processing **56** (1994), no. 6, 462–478.
16. F. Leymarie and M. D. Levine, *Simulating the grassfire transform using an active contour model*, IEEE PAMI **14** (1992), no. 1, 56–75.
17. Alan Liu, Elizabeth Bullitt, and Stephen. M. Pizer, *3d/2d registration via skeletal near projective invariance in tubular objects*, MICCAI'98, 1998, pp. 952–963.
18. Gregoire Malandain, Gilles Bertrand, and Nicholas Ayache, *Topological segmentation of discrete surfaces*, Int. Journal of Computer Vision **10** (1993), no. 2, 183–197.
19. Gregoire Malandain and Sara Fernandez-Vidal, *Euclidean skeletons*, Image and Vision Computing **16** (1998), 317–327.
20. Antoine Manzanera, Thierry M. Bernard, Francoise Preteux, and Bernard Longuet, *Medial faces from a concise 3d thinning algorithm*, ICCV'99 (Kerkyra, Greece), September 1999, pp. 337–343.
21. M. Näf, O. Kübler, R. Kikinis, M. E. Shenton, and G. Székely, *Characterization and recognition of 3d organ shape in medical image analysis using skeletonization*, IEEE Workshop on Mathematical Methods in Biomedical Image Analysis, 1996.
22. R. Ogniewicz, *Discrete voronoi skeletons*, Hartung-Gorre, 1993.
23. Lawrence Perko, *Differential equations and dynamical systems*, Springer-Verlag, 1986.

24. S. M. Pizer, A. Thall, and D. T. Chen, *M-reps: A new object representation for graphics*, ACM Transactions on Graphics **submitted** (1999).
25. Chris Pudney, *Distance-ordered homotopic thinning: A skeletonization algorithm for 3d digital images*, Computer Vision and Image Understanding **72** (1998), no. 3, 404–413.
26. M. Schmitt, *Some examples of algorithms analysis in computational geometry by means of mathematical morphology techniques*, Lecture Notes in Computer Science, Geometry and Robotics (J.D. Boissonnat and J.P. Laumond, eds.), vol. 391, Springer-Verlag, 1989, pp. 225–246.
27. T. B. Sebastian, H. Tek, J. J. Crisco, S. W. Wolfe, and B. B. Kimia, *Segmentation of carpal bones from 3d ct images using skeletally coupled deformable models*, MICCAI'98, 1998, pp. 1184–1194.
28. Damian J. Sheehy, Cecil G. Armstrong, and Desmond J. Robinson, *Shape description by medial surface construction*, IEEE Transactions on Visualization and Computer Graphics **2** (1996), no. 1, 62–72.
29. Evan C. Sherbrooke, Nicholas Patrikalakis, and Erik Brisson, *An algorithm for the medial axis transform of 3d polyhedral solids*, IEEE Transactions on Visualization and Computer Graphics **2** (1996), no. 1, 44–61.
30. Kaleem Siddiqi, Sylvain Bouix, Allen Tannenbaum, and Steven W. Zucker, *The hamilton-jacobi skeleton*, ICCV'99 (Kerkyra, Greece), September 1999, pp. 828–834.
31. G. D. Stetten and Stephen M. Pizer, *Automated identification and measurement of objects via populations of medial primitives, with application to real time 3d echocardiography*, IPMI'99, 1999, pp. 84–97.
32. Marek Teichmann and Seth Teller, *Assisted articulation of closed polygonal models*, 9th Eurographics Workshop on Animation and Simulation, 1998.
33. H. Tek and B. B. Kimia, *Symmetry maps of free-form curve segments via wave propagation*, ICCV'99 (Kerkyra, Greece), September 1999, pp. 362–369.
34. Yong Zhou, Aerie Kaufman, and Arthur W. Toga, *3d skeleton and centerline generation based on an approximate minimum distance field*, Int. Journal of the Visual Computer **14** (1998), no. 7, 303–314.

McGill University, School of Computer Science &, Center for Intelligent Machines, 3480 University Street, Montreal, QC H3A 2A7, Canada
*E-mail address*: sbouix@cim.mcgill.ca

McGill University, School of Computer Science &, Center for Intelligent Machines, 3480 University Street, Montreal, QC H3A 2A7, Canada
*E-mail address*: siddiqi@cim.mcgill.ca

DIMACS Series in Discrete Mathematics
and Theoretical Computer Science
Volume **55**, 2000

# A Mixed Variable Optimization Approach to Non-Rigid Image Registration

Anand Rangarajan and Haili Chui

ABSTRACT. Inter-subject non-rigid registration of cortical anatomical structures as seen in magnetic resonance images (MRI) is a challenging problem. The variability of the sulcal and gyral patterns across patients makes the task of registration especially difficult regardless of whether voxel- or feature-based techniques are used. In this paper, we present an approach to matching sulcal point features that have been interactively extracted by a neuroanatomical expert. The automated matching of the extracted sulci is accomplished by the robust point matching (RPM) algorithm. RPM is used to match two sets of labeled (or partially labeled) sulci. RPM iteratively solves for a thin-plate spline spatial mapping and the correspondences between the two sulcal point sets. Embedded within deterministic annealing, RPM is able to avoid poor, local minima while rejecting a fraction of the points in either set as outliers. Finally, we present synthetic benchmarks, with comparisons to the iterated closest point (ICP) algorithm on a Chinese character template and real-world comparisons between Talairach, mutual information-based matching and RPM on 5 subject MR images.

## 1. Introduction

Mixed (continuous and combinatorial) variable objective functions arise in many contexts. Such objective functions are appropriate modeling tools when the variables of interest comprise both analog parameters and discrete decision variables. In this paper, we use a mixed variable objective function formulation for a particular task—non-rigid medical image registration. In particular, we argue that the mixed variable objective function formulation is especially well suited for this problem.

In feature-based non-rigid medical image registration, we are asked to determine the spatial mapping that brings two sets of (usually) unlabeled features into register. In the most difficult version of this problem, the features are unlabeled and *sparse* and there is no prior knowledge of the set of correspondences between the two sets of features. Due to the vagaries of feature extraction (which we will set aside as a separate medical image analysis problem), a (hopefully small) subset of features in one set may have no corresponding features in the other and vice versa. Consequently, in the bare-bones version of non-rigid feature-based image

2000 *Mathematics Subject Classification.* Primary 49N99.

The authors were partially supported by NSF grant IIS-9906081.

registration, we have to determine the point-to-point correspondences between two sets of features (in 2D or in 3D), reject a subset of features in both sets as outliers and determine a spatial mapping that brings the two sets into register. Additional information in the form of feature attributes and shape primitives (curvature and torsion) is sometimes available and can be incorporated in the matching engine.

We formulate non-rigid registration as a mixed variable optimization problem. Since the spatial mapping is a space-to-space warp, a rigid, similarity, or affine transformation will not suffice. Instead, we model the spatial mapping as a high dimensional spline-based parametric mapping that can warp one set of *landmarks* onto another [**Boo89**]. The extent of the warp is directly proportional to the bending energy of the spline. Since we only have sparse feature information, the spline smoothly interpolates the warping in locations where there is no feature information. The spline is regularized in order to produce smooth mappings: greater the regularization, smaller the bending and vice versa [**Wah90**]. The resulting objective function on the spline parameters is a straightforward least-squares problem.

The spline parameters contribute to the analog side of the mixed variable objective function with the unknown point-to-point feature correspondences being their digital counterparts. While the spline is capable of modeling the mapping between the two sets of features, it requires the point-to-point correspondence information in order to get started. We model the correspondence problem using a linear assignment (or weighted bipartite matching) objective function [**PS82**]. The principal difference from standard linear assignment is the presence of outliers: a small subset of features may have no correspondences. In the standard linear assignment problem, the assignment variables form a permutation matrix. (A permutation matrix is a square, binary-valued matrix with its rows and columns summing to one.) Here, we add an extra row and column of binary outlier variables. All rows and columns still sum to one with the exception of the outlier row and column. If any variable with row index $i$ and column index $j$ (not in the outlier row and column) is one, it triggers a correspondence between feature $i$ in one set and feature $j$ in the other. If an outlier variable is one, that feature has no correspondence in the other set. With correspondence parameterized in this manner, we have fully specified both the analog spatial mapping and discrete correspondence variables.

Having formulated the mixed variable non-rigid matching objective, we now turn our attention to the optimization algorithm. First, we note that in the mixed variable objective formulation, the spatial mapping spline parameters appear as a conventional least-squares term (with the correspondences held fixed). Consequently, for *any fixed setting* of the correspondences, we can solve, in closed-form for the spatial mapping parameters. (As an interesting aside, the resulting objective function after elimination of the analog spatial mapping parameters is strikingly similar to the well-known quadratic assignment problem [**PW94**].) Therefore, the solution of the mapping parameters is straightforward, given the correspondences. What about the opposite direction? We have already mentioned that we have to solve a linear assignment problem for the correspondences. This is true for any setting of the spatial mapping parameters.

We have found it to be inefficient to exactly solve for linear assignments while keeping the spatial mapping parameters fixed. However, the above statement is merely based on experience rather than on firm principles. It turns out that the mapping parameters fluctuate a great deal when optimal solutions of the linear assignment problem are used. Instead, our heuristic is to co-vary the warping regularization parameter with the degree of *fuzziness* of the correspondences. The correspondence are made fuzzy by relaxing the binary constraint on the correspondences. Rather than search for a permutation matrix (with binary outlier rejection), we search for an optimum doubly substochastic matrix. (A doubly substochastic matrix is a matrix with positive entries and with all rows and columns summing at most to one [**Bha96**].) When the outlier row and column are used, we end up searching for a non-square matrix with positive entries and all rows and columns summing to one except for the outlier row and column. It is natural to ask the following question at this juncture. What is the relationship between the relaxed problem and the original linear assignment problem? Due to a simple extension of the Birkhoff-von Neumann theorem, we know that the set of doubly substochastic matrices form the convex hull of the set of permutations and binary outliers. Consequently, even though we have relaxed the search to doubly substochastic matrices, we are guaranteed in the limit to end up with permutations and binary outliers.

Our approach to solving the linear assignment problem using fuzzy doubly stochastic matrices as intermediate points is not new. Essentially, the pioneering credit goes to [**KY94**]. In [**KY94**], an extremely simple algorithm for linear assignments was introduced. Given a linear assignment benefit matrix, a doubly stochastic fuzzy matrix is obtained by exponentiating each benefit matrix entry (with the degree of fuzziness depending on an external *temperature* parameter). The row and column constraints are solved by iterated row and column normalization. This approach is guaranteed to converge to a doubly stochastic matrix. We denote this useful operation—exponentiation followed by iterated row and column normalization—the *softassign* [**GRL$^+$98, Sin64**]. In [**KY94**], it was shown that by gradually lowering the temperature parameter in the softassign, the optimal solution to linear assignments results. The process is completely deterministic save for a small amount of randomization to encourage symmetry breaking. We have adapted this approach to solve for fuzzy correspondences. It is trivial to extend the softassign to account for an extra outlier row and column. Having done this, we now have an optimal fuzzy doubly substochastic matrix for each setting of the spatial mapping parameters. As mentioned previously, the degree of regularization is co-varied with the temperature parameter. As the temperature is lowered, so is the regularization parameter resulting in progressively larger bending and warps. The final algorithm—called the *robust point matching* (RPM) algorithm—is an alternating optimization algorithm which is guaranteed to find a local optimum of the original mixed variable objective function.

We now turn to the particular medical image registration problem that is of interest to us. The recent development of brain imaging technologies (PET, MRI, fMRI) has provided rich information on the human brain. A potentially fruitful emerging area of research is human brain mapping [**TM96**] which requires a comprehensive statistical analysis of brain structure and function across diverse populations and different imaging modalities. A major requirement in brain mapping is that the imaging data from different subjects and modalities have to be placed in a common reference frame. Recent efforts have focused on using anatomical MR as the basis for such registration. After obtaining a common reference frame by performing *inter-subject* non-rigid anatomical MR registration, the data from other modalities can be placed in the common reference and then statistically analyzed.

Inter-subject anatomical registration is a difficult task due to the complexity and variability of brain structures. This is most obvious in the cortical regions. The folding of the cortical surfaces—the sulci and gyri—vary dramatically from person to person and, in some cases [**TM96**] are not even always present in each subject. However, the folding pattern is not completely arbitrary. In fact, the sulci often serve as important cortical landmarks. Furthermore, many cortical areas have been associated with critical brain functionalities (vision, language, motor control etc.) with the sulci often representing important functional boundaries. Cortical registration, despite its enormous difficulty, is hence highly desirable as a basis for further statistical quantitative analysis.

Our approach is based on matching feature points representing the sulcal structures. The points were obtained using a tool [**RZS$^+$98**] which allows a neuroanatomical expert to interactively trace sulci on a 3D skull-stripped MRI brain volume. In contrast to choosing a few landmarks, the tool allows us to represent sulci using hundreds of 3D points. Also, major sulci can be identified and easily labeled. It might seem that 3D space curves would be a better representation of the sulci traced by the expert. This turns out not to be true. Due to the length and the undulating characteristics of some of the sulci, it is usually necessary for the expert to break up a sulcus into a number of partially overlapping *traces*. Consequently, the set of traces corresponding to a sulcus is *more conveniently represented* as a 3D point-set than as a 3D space curve. We then match two sets of partially labeled sulci (extracted from two subjects' MRI) using the aforementioned RPM algorithm. While RPM has been previously developed and used for 2D rigid alignment [**RCM$^+$97**] and 2D and 3D affine warping [**GRL$^+$98, CRD$^+$99**], for the first time, we have developed the technique for 3D non-rigid warping. Since some sulcal structures in one subject may not have corresponding homologies in the other, RPM is able to reject a fraction of such non-homologies as outliers. Thin-plate splines are used for the spatial mapping. RPM returns both the point-to-point correspondences (and outliers) and the thin-plate spline mapping. We show results on five subject MR sulcal data and compare the results against a mutual-information-based intensity matching approach and our own previous approaches using RPM with affine and piecewise-affine spatial mappings.

## 2. Review

There are two principal approaches to non-rigid brain registration. Most existing methods can be divided into two categories; voxel-based and feature-based [**VPV93**].

Voxel-based approaches try to find the optimal transformation such that a local image intensity similarity measure is maximized. Most methods in this class allow highly complex transformations which are normally proportional to the size of the volume. Elastic media models, viscous fluid models [**CJM97**] or local smoothness models [**Gee95, CHPE95**] are introduced as constraints to guide the non-rigid spatial mapping. From these efforts, the need for non-rigid transformations is by now quite clear. Note, however that these algorithms are driven by local voxel intensities. Each voxel is treated equally without taking advantage of higher level geometric information (such as the sulcal and gyral patterns used here). In these methods, further anatomic validation is necessary to ensure that homologous sulci are indeed matched. Aware of this lack, landmark points was used as an initial step in [**CJM97**]. As Collins *et al.* [**CGE98**] point out in their recent work, the voxel intensity approach worked well for deep subcortical structures, but sometimes had difficulty aligning sulci and gyri. To correct this, in their recent work [**CGE98**], they used a chamfer distance measure [**Bor88**] on sulci points and combined it with their former voxel-based matching (ANIMAL) framework. All of these efforts are attempting to incorporate neuroanatomic geometric feature expertise into their registration engines.

Feature-based methods, as the name implies, capitalize on the information from different identifiable brain structures. Features which represent important brain structures are extracted. The features run the gamut of landmark points [**Boo89**], curves [**SL97**] or surfaces [**TT96, Dav97**]. Subsequently, these methods attempt to solve for the correspondence and transformation between the features. The spatial transformations resulting from feature matching are then propagated to the whole volume. Underlying the philosophy of feature matching is that homologous features always provide an effective anchor for registration. However, enthusiasm for these methods is usually tempered not only by the difficulty of feature extraction but also by the difficulty of simultaneously determining the correspondences or homologies and the spatial mapping. The first problem—feature extraction—usually calls for some residual manual intervention while the second problem—automated matching—involves the computationally demanding task of determining the correspondences and the spatial mapping. As our method basically belongs to this category, we discuss previous methods in some detail and compare them to ours.

Bookstein [**Boo89**] pioneered the usage of landmark points for registration and shape analysis. The thin-plate-spline is used as the spatial mapping between the two landmark point sets to generate a elastic transformation in which the bending energy is minimized. Since it basically relies on a few landmark points, the accuracy of their positioning is essential. The homologies between all landmark points is deemed known (in advance). In contrast, in our approach, the correspondences and the spatial mapping are co-determined from hundreds of sulci feature points. In addition, the anatomical variability between subjects can create many outliers, i.e., sulci points which do not match. Since we are using hundreds of points to represent the structural information, it is statistically much more robust and the noise or point "jitter" which may be caused by various sources such as the tracing process or sampling error, should not significantly affect the final result.

In Thompson *et al.*'s work [**TT96**], 3D active surfaces are used to extract the surfaces of lateral ventricle and outer cortex which are developmentally fundamental for the brain. An initial surface is first constructed from some fiducial points and is then relaxed towards the edges until a final balance is reached between the edge attraction force and the surface smoothness measure. To better represent the deep cortical structures (sulci), parametric mesh surfaces are also interactively extracted. A point-to-point mapping between the two *surfaces* is then calculated and a linearly weighted 3D volumetric warping is generated. Davatzikos [**Dav97**] has a similar framework where surface curvature maps at different scales are used for different brain structures. More consideration is given to the inhomogeneity within the brain. A more sophisticated elasticity model makes the algorithm more flexible at the ventricles and more powerful to account for some abnormal cases where, for example, tumors are involved. Both methods emphasize the importance of sulcal alignment and not surprisingly, the validation in [**TT96**] has shown that anatomically homogeneous points can be accurately aligned. As our method is based on matching cortical structures, it is quite similar to both of these approaches. However, we use point-sets as a representation for the sulci rather than surfaces. The major sulci are labeled which imposes strong constraints on the matching. Moreover, the non-rigid matching of 3D surfaces (parameterized by surface normals for example) is a difficult problem. The parameterization of cortical structures as partially labeled point-sets allows us to easily utilize Procrustes methods of shape analysis [**Goo91, RCB97**] (by equating the atlas with the Procrustes mean). Eigenanalysis of the error covariance matrix (around the Procrustes mean atlas) also yields valuable information regarding the dominant modes of deformation present in a population [**CTCG95**].

We have presented a detailed review of competing approaches to solving point correspondence problems elsewhere [**RCM+97**]. Here we briefly discuss chamfer distances [**Bor88, CGE98**] and the iterated closest point (ICP) matching algorithm [**BM92**]. The chamfer distance has been used in cortical registration by Sandor and Leahy [**SL97**] and Collins *et al.* [**CGE98**]. The main problem with the chamfer distance is that it uses a brittle nearest neighbor measure to assign correspondence. Nearest neighbor methods used in chamfer matching and ICP are problematic in the vicinity of outliers since they generate local minima. Unlike the chamfer matching in [**CGE98**] where a distance image is calculated from the Euclidean distance from each voxel to its nearest sulci point feature, we directly use the sulci point feature locations for the matching. Finally, we should mention the work presented in [**LvC98**] where a maximum clique approach is taken to matching relating sulcal representations. Maximum clique is a very difficult NP-complete problem [**GJ79**] which in this case increases the likelihood of getting stuck in local minima. Also, it is difficult to explicitly model non-rigid spatial mappings in the maximum clique approach [**LvC98**]. Consequently, the "engine" that does the work has to be pure sulcal correspondences making the problem more difficult.

## 3. A Non-Rigid Mapping Energy Function

First, we review thin-plate splines with an eye toward integrating it with our softassign correspondence engine. Later, we develop the full-blown non-rigid mapping and correspondence energy function from which we derive the TPS-RPM algorithm.

**3.1. Thin-Plate Splines.** Assume we have two sets of of *corresponding points* (in either 2D or 3D) and denote them as $\{u_i,\ i = 1, 2, \dots N\}$ and $\{v_i, i = 1, 2, \dots N\}$. For the sake of exposition, we present the 2D formulation. Each $u_i$ represents a point location with the homogeneous coordinate notation, i.e., $u_i = (1, u_{ix}, u_{iy})$. If we denote the dimensionality as $D$ ($D = 2$ for 2D here), $u_i$ is a $(D + 1)$ sized vector. The thin-plate spline fits a mapping function $f(u_i)$ between corresponding point-sets $u$ and $v$ by minimizing the following energy function:

$$
\begin{aligned}
E_{\text{tps}}(f) = \sum_{i=1}^{N} ||v_i - f(u_i)||^2 \\
+\lambda \int_{-\infty}^{+\infty}\int_{-\infty}^{+\infty} [(\tfrac{\partial^2 f}{\partial x^2})^2 + 2(\tfrac{\partial^2 f}{\partial x \partial y})^2 + (\tfrac{\partial^2 f}{\partial x^2})^2] dxdy
\end{aligned}
\tag{1}
$$

The first point-set $\{u_i\}$ is transformed as close as possible to the second point-set $\{v_i\}$ by minimizing the first error measurement term. Since there are an infinite number of mappings $f$ that can minimize the first term, the second term come into play to put constraints on the mapping. The second term is essentially a smoothness constraint on the mapping. The extent of the warping depends on the setting of the regularization parameter $\lambda$.

It is known that for this energy function with fixed $\lambda$, there exists a unique minimizing function $f$ specified by two parameter matrices $d$ and $w$:

$$f(u) = u \cdot d + \phi(u) \cdot w \tag{2}$$

where $d$ is an $(D+1) \times (D+1)$ affine transformation matrix, $w$ is a $N \times (D+1)$ warping coefficient matrix and $u$ is now any point location in 2D in homogeneous coordinates. The vector $\phi(u)$ is related to the thin-plate spline *kernel*. It is a $1 \times N$ vector for any $u$ where each entry $\phi_j(u) = c||u - u_j||^2 \log ||u - u_j||$ with $c$ being a constant [**Wah90**]. Loosely speaking, the thin-plate spline kernel contains the information about the point-set's internal structural relationships and generates a non-rigid warp when combined with the warping coefficient matrix $w$ (as shown above).

A special characteristic of the thin-plate spline is that the resulting transformation is always decomposed into a global affine transformation and a local non-affine warping component. The smoothness term above is a function of the warping components alone. This is a desirable property since the rotation, translation and global shear components (which are included in the affine transformation) are then not penalized. When the solution for $f$ in (2) is substituted into the TPS energy function in (1), we get

$$E_{\text{tps}}(d, w) = ||V - Ud - \Phi w||^2 + \lambda \, \text{trace}(w^T \Phi w) \tag{3}$$

where $U$ and $V$ are just concatenated versions of the point coordinates $u_i$ and $v_i$, and $\Phi$ is an $(N \times N)$ matrix formed from the $\phi(u_i)$. Each row of the newly formed matrix comes from the original vectors. The progression from (1) to (3) mirrors a transition from the standard TPS functional to an energy function defined on the affine $d$ and warping $w$ parameters. The least-squares objective function in (3) is minimized using a QR decomposition [**Wah90**].

$$U = [Q_1 Q_2] \begin{pmatrix} R \\ 0 \end{pmatrix} \tag{4}$$

where $Q_1$ and $Q_2$ are $N \times (D+1)$ and $N \times (N-D-1)$ orthonormal matrices, respectively. The matrix $R$ is upper triangular. With the QR decomposition in place, (3) becomes,

(5)
$$E_{tps}(\gamma, d) = ||Q_2^T V - Q_2^T \Phi Q_2 \gamma||^2 + ||Q_1^T V - Rd - Q_1^T \Phi Q_2 \gamma||^2 + \lambda \gamma^T Q_2^T \Phi Q_2 \gamma$$

where $w = Q_2\gamma$ and $\gamma$ is a $(N-D-1) \times (D+1)$ matrix. Setting $w = Q_2\gamma$ (which in turn implies that $U^T w = 0$) enables us to cleanly separate the warping into affine and non-affine subspaces.

The least-squares energy function in (5) can be first minimized w.r.t $\gamma$ and then w.r.t. $d$. The final solution for $w$ and $d$ are,

$$\widehat{w} = Q_2 \widehat{\gamma} = Q_2 (Q_2^T \Phi Q_2 + \lambda I_{(N-D-1)})^{-1} Q_2^T V \tag{6}$$

$$\widehat{d} = R^{-1}(Q_1^T V - \Phi \widehat{w}) \tag{7}$$

We call the minimum value of the TPS energy function obtained at the optimum $(\widehat{w}, \widehat{d})$ the "bending energy".

$$E_{bending} = \lambda \operatorname{trace}\left[V^T Q_2 (Q_2^T \Phi Q_2 + \lambda I_{(N-D-1)})^{-1} Q_2^T V\right] \tag{8}$$

**3.2. A Mixed Variable Objective Function.** We have shown above that once we have two sets of corresponding points, we can fit a thin-plate spline interpolant between them. We now take up the question of what happens when we do not know the correspondence between the two point-sets.

An interesting feature of the point matching problem (as formulated here) is that even though the simultaneous solution of mapping and correspondence is quite difficult, solving for one factor while keeping the other fixed is trivial. Given the correspondence, finding the optimal transformation is a simple least-squares problem as seen from (3) above. Given the non-rigid mapping, finding the correspondence between the point-sets is accomplished by solving a linear assignment problem [**GRL+98**]. Inspired by this fact, a lot of algorithms have tried to incorporate an update scheme wherein the mappings and the correspondences are alternatively optimized while keeping the other fixed [**Wel97, FA96**]. Our algorithm follow this basic idea. By alternating the update of the mapping and correspondence parameters, it is expected that the two solutions mutually improve one another during the process and finally converge to a reasonable, albeit sub-optimal solution.

Denote the correspondence parameter by the matrix $M$. The inner $n_1 \times n_2$ part of $M$ carries the correspondence information between the two point sets $\{u_i\}$ and $\{v_j\}$. When point $u_i$ matches $v_j$, $M_{ij}$ is equal to 1. Otherwise it is zero. To ensure one-to-one correspondence, each row and each column of $M$ should sum up to one. We also put in an extra row and an extra column in $M$ to take care of the outliers so that the row and column sum constraints still hold after including the outliers. The energy function with both the correspondence $M$ and the thin-plate spline $(d, w)$ is the following:

$$\begin{aligned} E(M, d, w) &= \sum_{i=1}^{n_1} \sum_{j=1}^{n_2} M_{ij} ||v_j - u_i \cdot d - \phi(u_i) \cdot w)||^2 \\ &+ \alpha(\sum_{i=1}^{n_1} M_{i(n_2+1)} + \sum_{j=1}^{n_2} M_{(n_1+1)j}) + \lambda \operatorname{trace}(w^T \Phi w) \end{aligned} \tag{9}$$

where $M$ with $(M_{ij} \in \{0,1\})$ always satisfy the constraints, $\sum_{i=1}^{n_1+1} M_{ij} = 1, \forall j \in \{1, \ldots, n_2\}$ and $\sum_{j=1}^{n_2+1} M_{ij} = 1, \forall i \in \{1, \ldots, n_1\}$. Note that the summation includes the outlier row and column. The cost function depends on two global parameters $\alpha$ and $\lambda$. The parameter $\alpha$ penalizes the outlier row and column. The parameter $\lambda$ controls the degree of bending of the thin-plate spline. The cost function in (9) contains two intertwined optimization problems: a standard least-squares problem w.r.t. the thin-plate spline spatial mapping $(d, w)$ and a (slightly modified) linear assignment problem on the correspondence matrix $M$.

**Optimization Heuristic:** We have elected to use an alternating algorithm approach to minimize the above cost function. Our algorithm alternates between the solutions to the linear assignment and least-squares sub-problems. As mentioned in the Introduction, we have found it inefficient to use optimal solutions to linear assignment at each step of the alternating procedure. Instead of solving the original, discrete, linear assignment problem, we solve a *fuzzy*, linear assignment problem. We assume that the correspondence matrix entries can take continuous, non-negative values. Our alternating update scheme is very simple. We start with a uniform initial value of $M$, with $d$ and $w$ set to zero. For the first step, we keep $M$ fixed and fit a thin-plate spline $(d, w)$ between $\{u_i\}$ and their currently corresponding points $\{v_i^{'} = \sum_{j=1}^{n_2} M_{ij} \cdot v_j\}$. (Note that we are not assuming correspondence. $v^{'}$ is obtained based on the current estimate of $M$.) At the second step, we keep $d$ and $w$ fixed and calculate the match matrix $M$ between the transformed $\{u_i' = f(u_i|d, w)\}$ and $\{v_j\}$ while making sure it always satisfies the row and column summation constraints. Steps one and two are repeated until the algorithm converges.

The theoretical underpinning of this approach comes from a straightforward extension to the well known Birkhoff-von Neumann theorem: "the set of doubly stochastic matrices is the convex hull of the set of permutation matrices." Since we have an outlier row and column, our correspondence matrix $M$ is not a permutation (though it contains one as a subset). We use the following extension to the Birkhoff-von Neumann theorem: "the set of doubly *substochastic* matrices is the convex hull of the set of matrices having at most one entry 1 in each row and each column and all other entries zero [**Bha96**]." The set of doubly substochastic matrices comprises matrices with nonnegative entries and with the sum of each row and each column being at most one.

Using this theorem, a fuzzy linear assignment cost function is constructed [**KY94**]. A barrier function (with an associated barrier function parameter) is used to keep the entries of $M$ non-negative. As the barrier function parameter is reduced to zero, the original, discrete linear assignment problem is approached. At each setting of the barrier function parameter, the solution to the fuzzy, linear assignment problem is a certain doubly substochastic matrix. In [**KY94**], it is shown that the optimal solution to linear assignment is obtained as the barrier function parameter approaches zero. In our case, the spatial mapping is updated at each step along with the solution to the fuzzy, linear assignment problem. Consequently, even though both solutions (spatial mapping and correspondence) are optimal at each setting of the barrier function parameter, the combination is not. Our approach is only guaranteed to find a local minimum for the mapping *and* the correspondence.

**3.3. The TPS-RPM Algorithm.** Putting everything together, the final energy function which is actually minimized by our algorithm is as follows:

$$\begin{aligned} E_{\text{final}}(M, d, w) &= \sum_{i=1}^{n_1}\sum_{j=1}^{n_2} M_{ij}||v_j - u_i \cdot d - \phi(u_i) \cdot w)||^2 \\ &+\alpha(\sum_{i=1}^{n_1} M_{i(n_2+1)} + \sum_{j=1}^{n_2} M_{(n_1+1)j}) + T\sum_{i=1}^{n_1}\sum_{j=1}^{n_2} M_{ij}\log M_{ij} \\ &+\lambda_1 \text{ trace}(w^T\Phi w) + \lambda_2 \text{ trace}[(d-I)^T(d-I)] \\ &+\sum_{i=1}^{n_1}\xi_i(\sum_{j=1}^{n_2} M_{ij} - 1) + \sum_{j=1}^{n_2}\eta_j(\sum_{i=1}^{n_1} M_{ij} - 1) \end{aligned} \tag{10}$$

We briefly go through all the components of the energy function. The first term is just the error measure term which becomes zero in the absence of any regularization. The second term with a parameter $\alpha$ is used to guard against nullmatches. We usually set it to a value close to zero. The third term is an $x \log x$ entropy barrier function with a temperature barrier function parameter $T$. The entropy barrier function ensures positivity of $M$. For more details regarding the choice of the barrier function, please see [**GRL**$^+$**98, KY94**]. The fourth term is the standard thin-plate spline regularization term which penalizes the local warping coefficients $w$. We also use the fifth term which is not usually employed in the thin-plate spline literature, to constrain the affine mapping $d$ by penalizing the residual part of $d$ which is different from a identity matrix $I$. The main reason for this term is to constrain the affine from unphysical reflection mappings which can flip the whole plane. The behavior of these two terms is controlled by two parameters $\lambda_1$ and $\lambda_2$. Inspired by the annealing schedule on the correspondence, we also use a linear annealing schedule ($\lambda_i = \lambda_i^{init} T$, $i = 1, 2$) for these two parameters instead of using a fixed value. To provide more freedom for the affine transformation, $\lambda_2^{init}$ is set to be much smaller than $\lambda_1^{init}$. The sixth and seventh terms are the row and column constraints rewritten using Lagrange parameters $\xi$ and $\eta$. We never directly solve for the Lagrange parameters. Instead, we use the softassign with its iterated row and column normalization (Sinkhorn) procedure [**Sin64**] which guarantees that $M$ is doubly substochastic. This is equivalent to using the following closed form solution for the matrix $M$:

$$M_{ij} = \exp[-\frac{(||v_j - u_i \cdot d - \phi(u_i) \cdot w||^2 - \alpha + \xi_i + \eta_j)}{T} - 1] \tag{11}$$

The Sinkhorn balancing technique of alternating row and column normalizations is used to satisfy the row and column constraints [**GRL**$^+$**98**]. The parameters $(d, w)$ are held fixed during this process. When the correspondences are held fixed, we use a least-squares approach to solve for $(d, w)$. This is implemented using a QR decomposition, along the lines presented above [**Wah90**].

We now explain the setup for the remaining parameters. The parameter $\lambda_1^{init}$ is set to 1 and $\lambda_2^{init}$ is set to $0.01 \cdot \lambda_1^{init}$. The alternating update of the correspondence $M$ and the transformation $d, w$ is repeated 5 times after which they are usually close to convergence. We then decrease the temperature $T$ and repeat the entire process. The thin-plate spline in 2D and in 3D are almost identical except that in 3D, a different kernel of the form $\phi_b(v_a) = ||v_b - v_a||$ is used [**Wah90**].

**The TPS-RPM Algorithm**
Initialize $M$, $d$ and $w$.
Initialize $\lambda_1$, $\lambda_2$ and $T$.
**Begin A: Deterministic Annealing**
    **Begin B: Relaxation.**
        Update $M$ by softassign.
        Weighted least-squares solution for the non-rigid mapping $(d, w)$.
    **End B**
    Decrease $T$.
    Decrease $\lambda_1, \lambda_2$.
**End A**

## 4. Experiments and Results:

The iterated closest point (ICP) algorithm [**BM92**] is actually quite similar to our algorithm. The main difference is that ICP depends on the nearest-neighbor heuristic and always returns a binary correspondence, which is not guaranteed to be one-to-one. However, it is close to the limiting case of our algorithm when annealing is replaced with quenching (starting $T$ close to zero). Since $T$ can be regarded as a search range parameter, it is not hard to see that the over-simplified treatment of correspondence in ICP makes it much more vulnerable to local minima. A second difference is that ICP relies on other heuristics to handle outliers. For example, outliers are rejected by setting a dynamic thresholding mechanism [**FA96**]. Distances between nearest point pairs are first fit to a Gaussian distribution. Points with distance measure values larger than the mean plus $L$ (usually set to 3) times the standard deviation are then rejected as outliers. We implemented this improved version of ICP as a benchmark comparison for our algorithm.

**4.1. Evaluation on Synthetic Examples.** To test RPM's performance, we ran a lot of experiments on synthetic data with different degrees of warping, different amounts of noise and different amounts of outliers and compared it with ICP.

After a *template* point-set is chosen, we apply a randomly generated non-rigid transformation to warp it. Then we add noise or outliers to the warped point-set to get a new *target* point-set. Instead of TPS, we use a different non-rigid mapping, namely Gaussian radial basis functions (RBF) [**YG89**] for the random transformation. The coefficients of the RBF were sampled from a Gaussian distribution with a zero mean and a standard deviation $s_1$. Increasing the value of $s_1$ generates more widely distributed RBF coefficients and hence leads to generally larger deformation. A certain percentage of outliers and random noise are added to the warped template to generate the target point-set. We then used both ICP and RPM to find the best TPS to map the template set onto the target set. The errors are computed as the mean squared distance between the warped template using the TPS found by the algorithms and the warped template using the ground truth Gaussian RBF.

We conducted three series of experiments. In the first series of experiments, the template was warped through progressively larger degrees of non-rigid warpings. The warped templates were used as the target data without adding noise or outliers. The purpose is to test the algorithms' performance on solving different degrees of deformations. In the second series, different amounts of Gaussian noise (standard deviation $s_2$ from 0 to 0.05) were added to the warped template to get the target data. A medium degree of warping was used to warp the template. The purpose is to test the algorithms' tolerance of noise. In the third series, different amounts of random outliers (outlier to original data ratio $s_3$ ranging from 0 to 2) were added to the warped template. Again, a medium degree of warping was used. 100 random experiments were repeated for each setting within each series.

The template used comes from a Chinese character (blessing), which is a rather complex pattern. All three series of experiments were run on this template.

Some examples of these experiments are shown in Figures 1, 2 and 3. The final statistics (error means and standard deviations for each setting) are shown in Figure 4. ICP's performance deteriorates much faster when the examples become harder due to any of these three factors—degree of deformation, amount of noise or amount of outliers. The difference is most evident in the case of outliers. ICP starts failing in most of the experiments and gives huge errors once outliers are added. On the other hand, RPM seems to be relatively unaffected. It is very interesting that it remains so even when the numbers of outliers far exceeds the number of true data points.

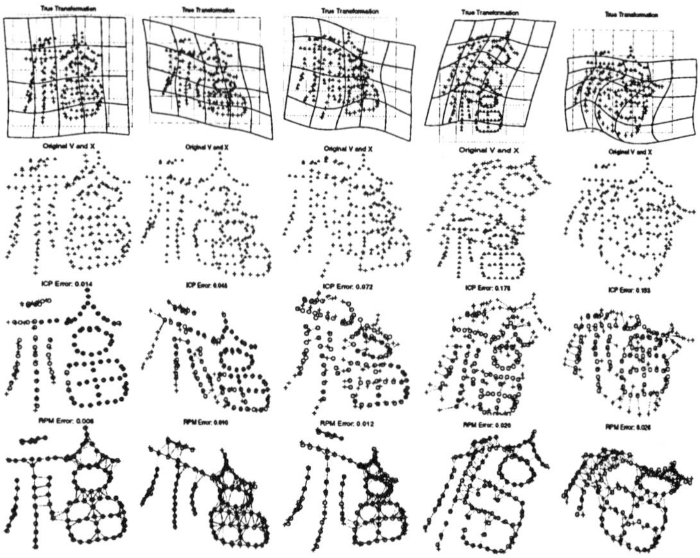

FIGURE 1. **Experiments on Deformation.** From left to right, increasing degree of deformation. Top Row: warped template. Second Row: template and target(same as the warped template). Third Row: ICP results. Bottom Row: RPM results.

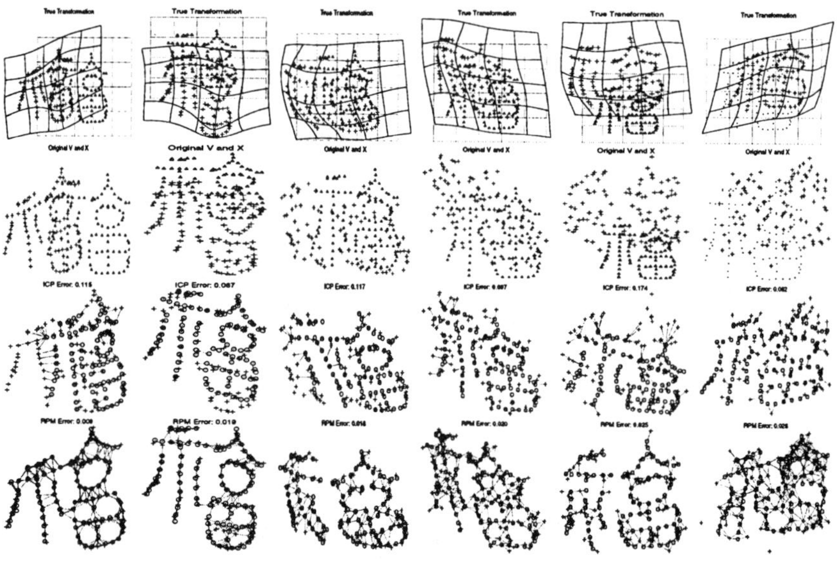

FIGURE 2. **Experiments on Noise.** From left to right, increasing amount of noise. Top Row: warped template. Second Row: template and target (warped template but with noise now). Third Row: ICP results. Bottom Row: RPM results.

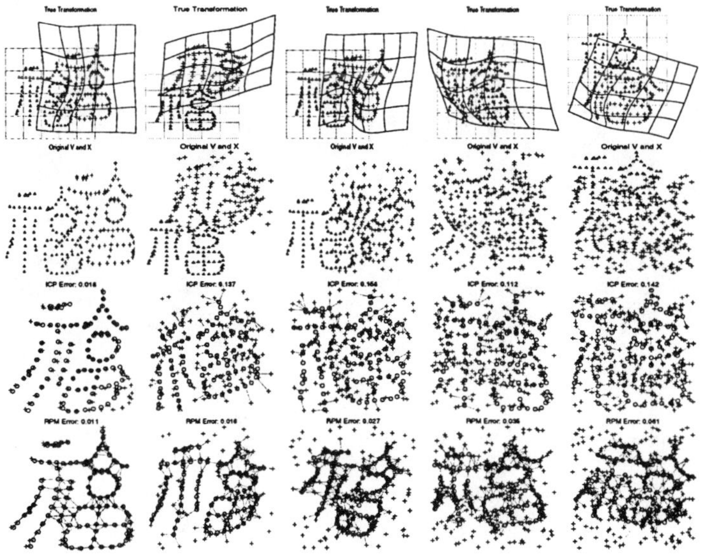

FIGURE 3. **Experiments on Outliers.** From left to right, increasing amount of outliers. Top Row: warped template. Second Row: template and target (warped template but with outliers now). Third Row: ICP results. Bottom Row: RPM results.

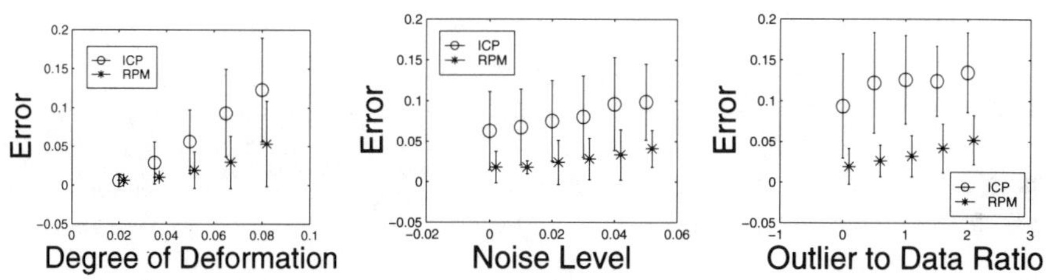


FIGURE 4. **Statistics of the Synthetic Experiments.** Results using the Chinese character template. Note especially how the RPM's errors increase much slower than ICP in all the cases.

**4.2. Experiments on Brain Mapping.** The sulcal tracing was done by on an SGI graphics platform [**RZS+98**] with a ray-casting technique that allows drawing in 3D space by projecting 2D coordinates of the tracing onto the exposed cortical surface. A screenshot of the tool is shown on the left in Figure 5.

The inter-hemispheric fissure and 10 other major sulci (superior frontal, central, post-central, Sylvian and superior temporal on both hemispheres) were extracted as point features. A sulcal point-set extracted from one subject is shown on the right in Figure 5.

The original sulcal point-sets normally contain around 3,000 points each. The point-set is first sub-sampled to have around 300 points by taking every tenth point. The original MRI volume's size is 106(X) x 75(Y) x 85(Z, slices). With that in mind, it is reasonable to assume that the average distances between points before registration should be in the range of 10 ~ 100. We set our starting temperature to be roughly in the same scale. After registration, we would expect the average

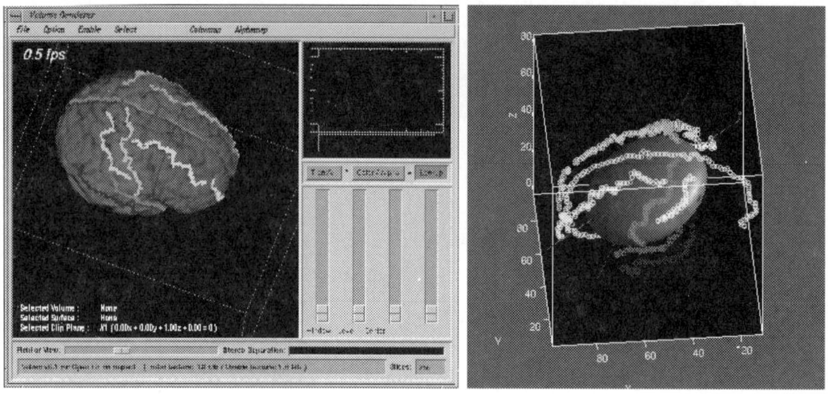


FIGURE 5. Left: A screenshot of the sulcal tracing tool with some traced sulci on the 3D MR brain volume.. Right: Sulci extracted and displayed as point-sets. Each sulcus is shown in a different color.

distance between corresponding points to be within a few voxels (say, 1 ~ 10). Our final temperature should be slightly smaller. From these considerations, we set the RPM annealing parameters to be the following: $T_{\text{init}} = 50$ , $T_{\text{final}} = 1$ , $T_{\text{anneal-rate}} = 0.95$ . The regularization parameter $\lambda_2$ is set to force the affine mapping to be small at first. We then decrease it by $\lambda_{\text{anneal-rate}} = 0.8$ at the end of every temperature iteration. As mentioned above, the idea is that the regularization should prevent the affine and the thin-plate mappings to be over determined by the initial fuzzy correspondence at first; once the algorithm starts moving towards the right correspondence, which usually happens within the first few iterations, the regularization should be relaxed by decreasing $\lambda_1$ and $\lambda_2$ with the temperature. Actually, we observed that normally any annealing rate between 0.7 and 0.9 works quite well.

We applied RPM to five sulcal point-sets and compared it with two other methods which also use affine (and piecewise-affine) transformations for brain registration. As we mentioned in the review section, we suspect that the voxel-based methods' performance would not be as satisfying as feature-based methods for sulcal alignment. To test this, we compares RPM with a voxel-based affine matching method [**Stu97**] which maximizes the mutual information between the two volumes.

By defining a common coordinate system, the Talairach method [**Tal67, TS88**] is a standard technique for brain alignment. A piecewise affine transformation is applied to 12 rectangular regions of brains defined by landmark points of anterior and posterior commissures and extrema of the cortex. We used the Talairach program available as part of the MEDx package (from Sensor Systems Inc.) to align 5 brains. Sulcal points were traced on the resulting brain volumes.

The volumes were then matched by the voxel-based method described in [**Stu97**] and the resulting spatial mapping was applied to the sulcal points. RPM was separately run on the sulcal point-sets, and both the result from a simple global affine transformation and the result from both global and piecewise affine transformations are shown in Figures 6 and 7. Since we register every brain to the first one, after registration, the minimum distance from each sulcal point in the current brain to

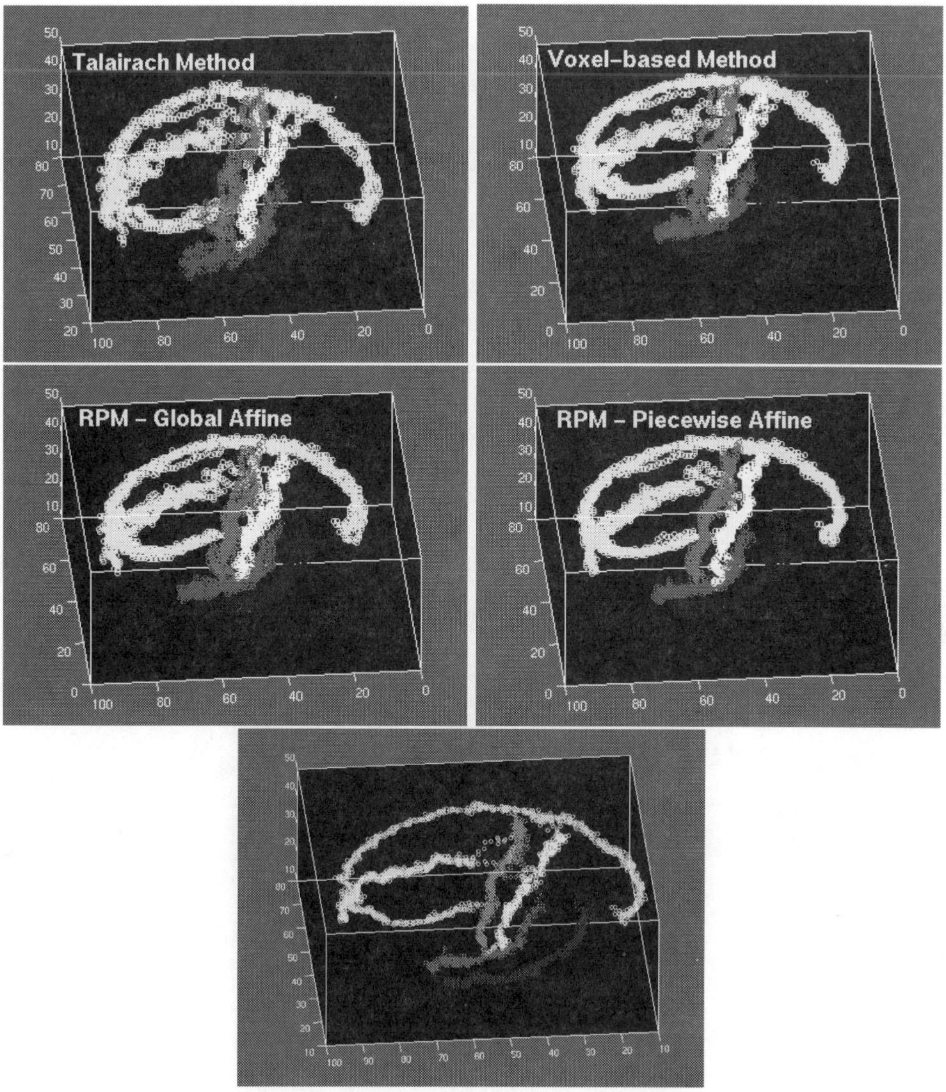


FIGURE 6. Sulcal points alignment. The alignment results of five brains' sulcal point-sets on the left side of the brain are shown together. Top row left: Talairach method. Top row right: The results of a voxel-based method by maximizing the mutual information between volume intensity with a global affine transformation. Middle row left: RPM with a global affine transformation. Middle row right, RPM with a piecewise affine transformation. Bottom row: RPM with a thin-plate spline transformation. We see the improvement of RPM over both Talairach and the voxel-based approach.

the first is calculated. The mean and variance of such minimum distances for each sulcus are calculated for quantitative comparison.

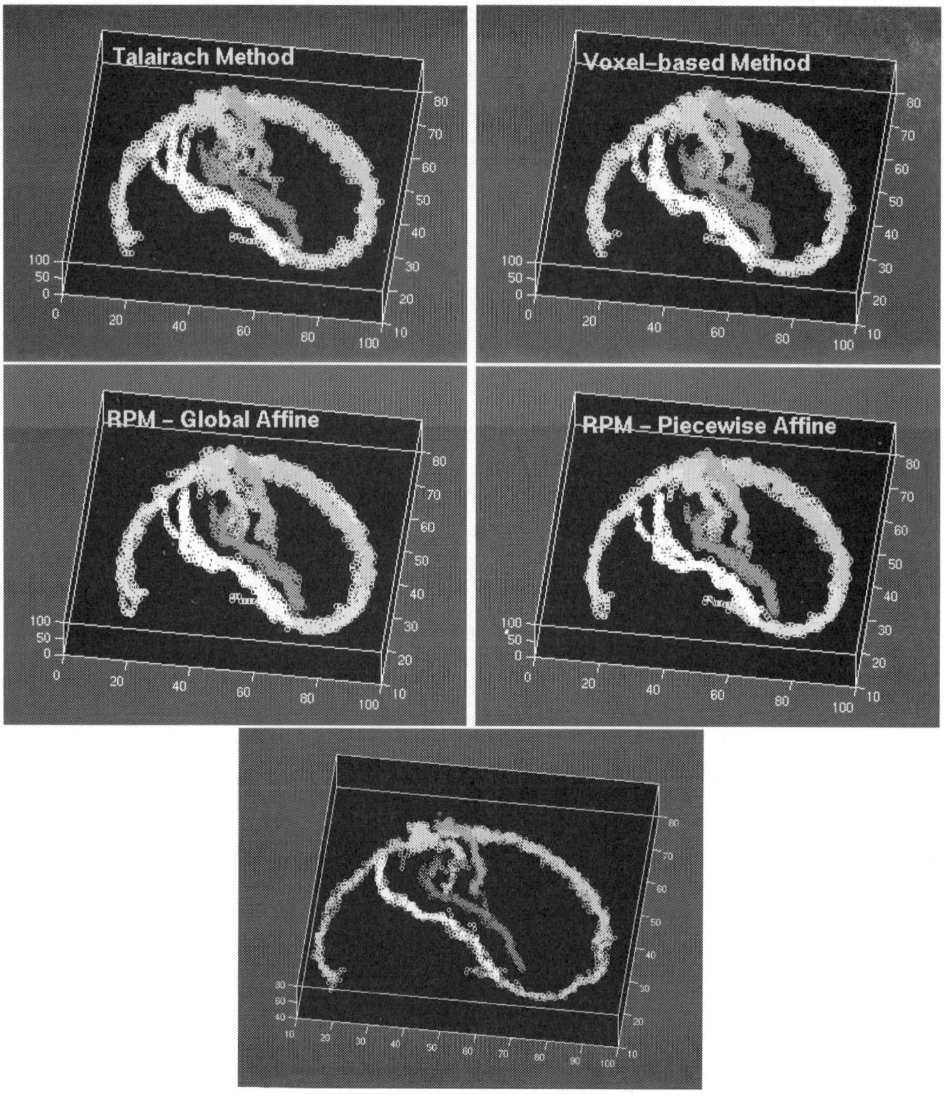


FIGURE 7. Sulcal points alignment. Same as the above except the data from the other side of the brain are shown.

The above comparison of Talairach with RPM and voxel-based approaches shows that RPM can significantly improve upon Talairach in most cases even though it may have less degrees of freedom. The voxel-based approach's performance is mixed; it gives bigger error for 5 of the 11 sulci. The significant improvement from the global affine transformations by allowing piecewise affine and thin-plate spline mappings confirmed our belief in the importance of non-rigid transformations.

## 5. Conclusions

The point matching problem become much harder when high dimensional non-rigid transformations are involved. From the experiments, we demonstrated that

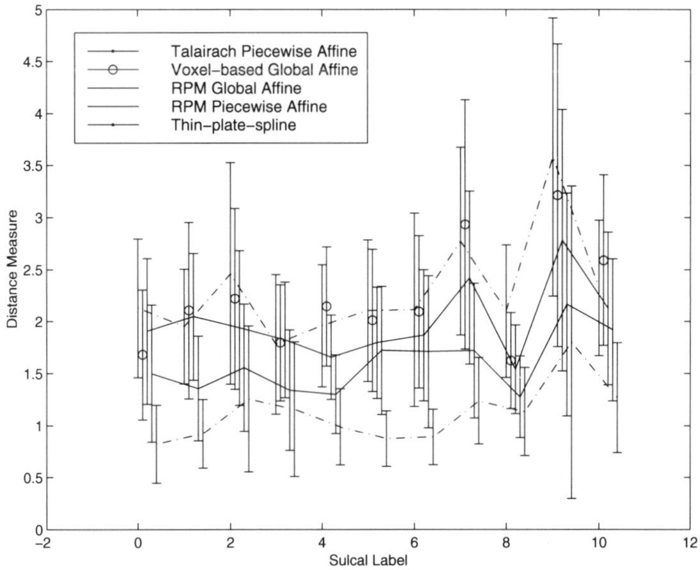


FIGURE 8. The minimum distance measure (as described in text) for each sulcus (sulcal label: 0 - interhemispheric fissure; 1,2 - central sulcus; 3,4 - Sylvian fissure; 5,6 - superior temporal sulcus; 7,8 - post-central sulcus; 9,10 - superior frontal sulcus.) The figure shows the comparison between Talairach , voxel-based method and RPM with global affine, piecewise affine and thin-plate spline mappings. The errors for the thin-plate spline mappings are the smallest.

our robust point matching (RPM) algorithm performs quite well even when the spatial mapping is a high dimensional thin-plate spline transformation. Our algorithm is able to solve non-rigid point matching problems with up to 100 points in less than 1 minute on an PentiumII 400MHz x86/linux machine (with the code implemented in MATLAB$^{\text{TM}}$). TPS-RPM is a fast and robust algorithm.

The kernel of our brain mapping algorithm is a mixed variable objective function. With the correspondence parameterized by a permutation matrix (and outliers) and the mapping by a thin-plate spline, we have essentially found a novel way to express a difficult non-rigid registration problem. After formulating it in this way, a particular alternating algorithm was used to solve for both the mapping and the correspondence. It turned out that solving for a fuzzy correspondence matrix went hand-in-hand with the solution for the non-rigid mapping. However, this alternating algorithm is just another heuristic device. It remains to be seen if there are other better and qualitatively different algorithms that can also minimize our mixed variable non-rigid registration objective function.

## Acknowledgements

This work was partially supported by an NSF grant to A.R. We thank Jim Rambo and Bob Schultz for assistance with the anatomical sulcal brain MRI data.

## References

[Bha96] R. Bhatia, *Matrix analysis*, Graduate Texts in Mathematics, vol. 169, Springer, New York, NY, 1996, page 38.

[BM92] P. J. Besl and N. D. McKay, *A method for registration of 3-D shapes*, IEEE Trans. Patt. Anal. Mach. Intell. **14** (1992), no. 2, 239–256.

[Boo89] F. L. Bookstein, *Principal warps: Thin-plate splines and the decomposition of deformations*, IEEE Trans. Patt. Anal. Mach. Intell. **11** (1989), no. 6, 567–585.

[Bor88] G. Borgefors, *Hierarchical chamfer matching: a parametric edge matching algorithm*, IEEE Trans. Patt. Anal. Mach. Intell. **10** (1988), no. 6, 849–865.

[CGE98] D. Collins, G. Goualher, and A. Evans, *Non-linear cerebral registration with sulcal constraints*, Medical Image Computing and Computer-Assisted Intervention (W. Wells, A. Colchester, and S. Delp, eds.), Lecture Notes in Computer Science, vol. 1496, Springer, 1998, pp. 974–984.

[CHPE95] D. Collins, C. Holmes, T. Peters, and A. Evans, *Automatic 3D model-based neuroanatomical segmentation*, Human Brain Mapping **3** (1995), no. 3, 190–208.

[CJM97] G. Christensen, S. Joshi, and M. Miller, *Volumetric transformation of brain anatomy*, IEEE Trans. Med. Imag. **16** (1997), no. 6, 864–877.

[CRD+99] H. Chui, J. Rambo, J. Duncan, R. Schultz, and A. Rangarajan, *Registration of cortical anatomical structures via robust 3D point matching*, Information Processing in Medical Imaging (IPMI '99), Springer, New York, 1999, pp. 168–181.

[CTCG95] T. Cootes, C. Taylor, D. Cooper, and J. Graham, *Active shape models: Their training and application*, Computer Vision and Image Understanding **61** (1995), no. 1, 38–59.

[Dav97] C. Davatzikos, *Spatial transformation and registration of brain images using elastically deformable models*, Computer Vision and Image Understanding: Special Issue on Medical Imaging **6** (1997), no. 2, 207–222.

[FA96] J. Feldmar and N. Ayache, *Rigid, affine and locally affine registration of free-form surfaces*, Intl. J. Computer Vision **18** (1996), no. 2, 99–119.

[Gee95] J. Gee, *Probabilistic matching of deformed images*, Ph.D. thesis, Dept. of Computer and Information Science, University of Pennsylvania, Philadelphia, PA, 1995.

[GJ79] M. R. Garey and D. S. Johnson, *Computers and intractability: a guide to the theory of np-completeness*, W. H. Freeman, San Francisco, CA, 1979.

[Goo91] C. Goodall, *Procrustes methods in the statistical analysis of shape*, J. R. Statist. Soc. B **53** (1991), no. 2, 285–339.

[GRL+98] S. Gold, A. Rangarajan, C. P. Lu, S. Pappu, and E. Mjolsness, *New algorithms for 2-D and 3-D point matching: pose estimation and correspondence*, Pattern Recognition **31** (1998), no. 8, 1019–1031.

[KY94] J. J. Kosowsky and A. L. Yuille, *The invisible hand algorithm: Solving the assignment problem with statistical physics*, Neural Networks **7** (1994), no. 3, 477–490.

[LvC98] G. Lohmann and D. von Cramon, *Sulcal basin and sulcal strings as new concepts for describing the human cortical topography*, Workshop on Biomedical Image Analysis, IEEE Press, June 1998, pp. 41–54.

[PS82] C. Papadimitriou and K. Steiglitz, *Combinatorial optimization*, Prentice-Hall, Inc., Englewood Cliffs, NJ, 1982.

[PW94] P. M. Pardalos and H. Wolkowicz (eds.), *Quadratic assignment and related problems*, DIMACS series in discrete mathematics and theoretical computer science ; vol 16., American Mathematical Society, Providence, R.I., 1994.

[RCB97] A. Rangarajan, H. Chui, and F. Bookstein, *The softassign Procrustes matching algorithm*, Information Processing in Medical Imaging (IPMI '97), Springer, 1997, pp. 29–42.

[RCM+97] A. Rangarajan, H. Chui, E. Mjolsness, S. Pappu, L. Davachi, P. Goldman-Rakic, and J. Duncan, *A robust point matching algorithm for autoradiograph alignment*, Medical Image Analysis **4(1)** (1997), 379–398.

[RZS+98] J. Rambo, X. Zeng, R. Schultz, L. Win, L. Staib, and J. Duncan, *Platform for visualization and measurement of gray matter volume and surface area within discrete cortical regions from MR images*, NeuroImage **7** (1998), no. 4, 795.

[Sin64] R. Sinkhorn, *A relationship between arbitrary positive matrices and doubly stochastic matrices*, Ann. Math. Statist. **35** (1964), 876–879.

[SL97] S. Sandor and R. Leahy, *Surface based labeling of cortical anatomy using a deformable atlas*, IEEE Trans. Med. Imag. **16** (1997), no. 1, 41–54.

[Stu97] C. Studholme, *Measures of 3D medical image alignment*, Ph.D. thesis, University of London, London, U.K., 1997.

[Tal67] J. Talairach, *Atlas d'anotomie stereotaxique du telencephale: etudes anatomo-radiologiques*, JMasson and Cie, Paris, 1967.

[TM96] A. Toga and J. Mazziotta, *Brain mapping: The methods*, Academic Press, 1996.

[TS88] J. Talairach and G. Szikla, *Co-planar stereotaxic atlas of the human brain*, Thieme Medical Pub., New York, 1988.

[TT96] P. Thompson and A. W. Toga, *A surface-based technique for warping three-dimensional images of the brain*, IEEE Trans. Med. Imag. **5** (1996), no. 4, 402–417.

[VPV93] P. Van den Elsen, E. Pol, and M. Viergever, *Medical image matching—A review with classification*, IEEE Engineering in Medicine and Biology (1993), 26–38.

[Wah90] G. Wahba, *Spline models for observational data*, SIAM, Philadelphia, PA, 1990.

[Wel97] W. Wells, *Statistical approaches to feature-based object recognition*, Intl. J. Computer Vision **21** (1997), no. 1/2, 63–98.

[YG89] A. L. Yuille and N. M. Grzywacz, *A mathematical analysis of the motion coherence theory*, Intl. J. Computer Vision **3** (1989), no. 2, 155–175.

Departments of Diagnostic Radiology and Electrical Engineering, Yale University
*E-mail address*: `anand@noodle.med.yale.edu`

Department of Electrical Engineering, Yale University
*E-mail address*: `chui@noodle.med.yale.edu`

DIMACS Series in Discrete Mathematics
and Theoretical Computer Science
Volume **55**, 2000

# On Computing the Nearest Neighbor Interchange Distance

Bhaskar DasGupta, Xin He, Tao Jiang, Ming Li, John Tromp, and Louxin Zhang

ABSTRACT. In the practice of molecular evolution, different phylogenetic trees for the same group of species are often produced either by procedures that use diverse optimality criteria [**24**] or from different genes [**15, 16, 17, 18, 14**]. Comparing these trees to find their *similarities* (*e.g.* agreement or consensus) and *dissimilarities*, *i.e. distance*, is thus an important issue in computational molecular biology. The *nearest neighbor interchange* (nni) distance [**29, 28, 34, 3, 6, 2, 19, 20, 23, 33, 22, 21, 26**] is a natural distance metric that has been extensively studied. Despite its many appealing aspects such as simplicity and sensitivity to tree topologies, computing this distance has remained very challenging, and many algorithmic and complexity issues about computing this distance have remained unresolved. This paper studies the complexity and efficient approximation algorithms for computing the nni distance and a natural extension of this distance on weighted phylogenies. The following results answer many open questions about the nni distance posed in the literature.

1. Computing the nni distance between two labeled trees is NP-complete. This solves a 25 year old open question appearing again and again in, for example, [**29, 34, 3, 6, 2, 19, 20, 23, 22, 21, 26**].

2. Computing the nni distance between two unlabeled trees is also NP-complete. This answers an open question in [**3**] for which an erroneous proof appeared in [**23**].

3. Biological applications motivate us to extend the nni distance to *weighted* phylogenies, where edge weights indicate the time-span of evolution along each edge. We present an $O(n^2)$ time approximation algorithm for computing the nni distance on weighted phylogenies with a performance ratio of $4 \log n + 4$, where $n$ is the number of leaves in the phylogenies.

We also observe that the nni distance is in fact identical to the *linear-cost subtree-transfer* distance on unweighted phylogenies discussed in [**4, 5**]. Some consequences of this observation are also discussed.

---

2000 *Mathematics Subject Classification.* Primary 68Q17, 68W40; Secondary 68Q25.

The results reported here also form a subset of the results that appeared in *Proc. 8th Annual ACM-SIAM Symposium on Discrete Algorithms*, 1997, pp. 427-436 [**4**]. The remaining results of the conference paper which do not appear in this paper appeared separately in *Algorithmica*, Vol. 25, No. 2, pp. 176-195, 1999.

The first author was supported by an CGAT (Canadian Genome Analysis and Technology) grant. The second author was supported in part by CGAT and NSF grant 9205982. The third author was supported in part by NSERC Operating Grant OGP0046613 and CGAT. The fourth author was supported by NSERC Operating Grant OGP0046506 and CGAT. The fifth author was supported by an NSERC International Fellowship and CGAT. .

Work done while the first author was at University of Waterloo and McMaster University, the second author was visiting at University of Waterloo, the third author was visiting University of Washington, and the fifth and the sixth authors were at University of Waterloo.

## 1. Introduction

The evolution history of organisms is often conveniently represented as trees, called *phylogenetic trees* or simply *phylogenies.* Such a tree has uniquely labeled leaves and unlabeled internal nodes, is either *unrooted* or *rooted* (if the evolutionary origin is known), and usually all of whose internal nodes have degree 3. Over the past few decades, many different objective criteria and algorithms for reconstructing phylogenies have been developed, including (not exhaustively) parsimony [**8, 11, 31**], compatibility [**25**], distance [**12, 30**], and maximum likelihood [**1, 8, 9**]. The outcomes of these methods usually depend on the data and the amount of computational resources applied. As a result, in practice they often lead to different trees on the same set of species [**24**]. It is thus of interest to compare phylogenies produced by different methods or by the same method on different data for similarity and discrepancy. Several metrics for measuring the distance between phylogenies have been proposed in the literature. Among these metrics, the best known is perhaps the *nearest neighbor interchange* (nni) distance introduced independently in [**28**] and [**29**].

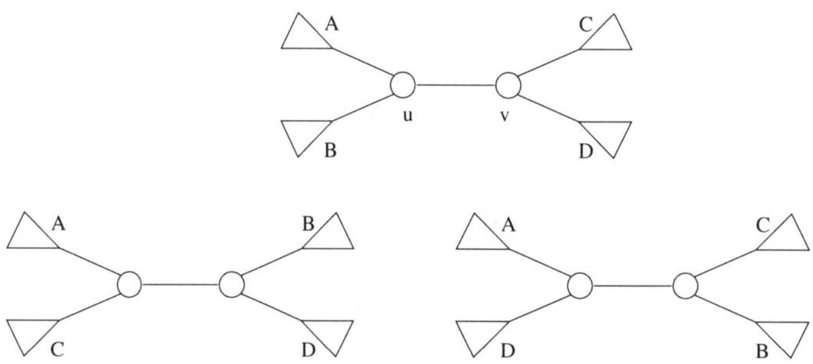


FIGURE 1. The two possible nni operations on an internal edge $(u, v)$: exchange $B \leftrightarrow C$ or $B \leftrightarrow D$.

An nni operation swaps two subtrees that are separated by an internal edge $(u, v)$, as shown in Figure 1. (An edge $(u, v)$ is *internal* if neither $u$ nor $v$ is a leaf.) The nni operation is said to *operate* or *perform* on this internal edge. The nni distance, $D_{nni}(T_1, T_2)$, between two trees $T_1$ and $T_2$ is defined as the minimum number of nni operations required to transform one tree into the other, as illustrated in Figure 2.

The complexity of computing the nni distance has been open for more than 25 years (since [**29**]). The problem is surprisingly subtle as witnessed by the history of many erroneous results, disproved conjectures and a faulty NP-completeness proof [**2, 19, 20, 23, 22, 26, 34**]. The question is open even for the simpler case where the trees are unlabeled. The faulty NP-completeness proof [**23**] we mentioned above was for this case.

A phylogeny may also have *weights* on its edges, where an edge weight (more commonly known as *branch length* in genetics) could represent the evolutionary distance along the edge. Many phylogeny reconstruction methods, including the distance and maximum likelihood methods, actually produce weighted phylogenies. Comparison of weighted phylogenies has recently been studied in [**24**]. The distance measure adopted is based on the difference in the partitions of the leaves induced

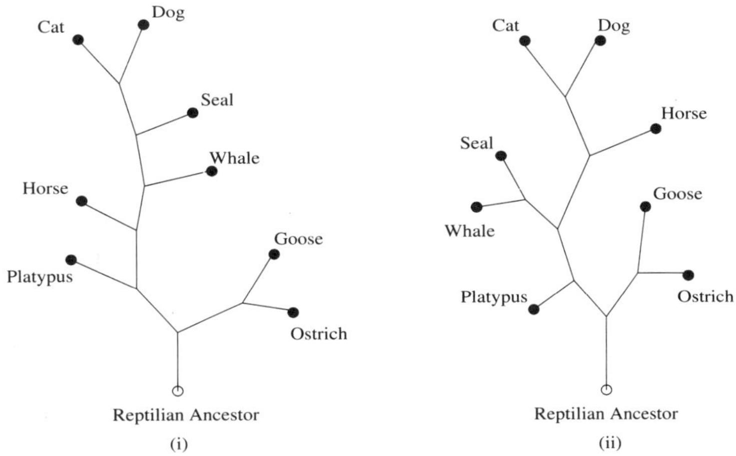


FIGURE 2. The nni distance between (i) and (ii) is 2.

by the edges in both trees, and has the drawback of being somewhat insensitive to the tree topologies [**10**]. The nni distance can be naturally extended to weighted phylogenies. An nni operation is simply charged a cost equal to the weight of the internal edge it operates on. Intuitively this extension of the nni distance is more sensitive to the tree topologies than the one in [**24**].

In this paper, we study the computational complexity and efficient approximation algorithms concerning the nni distance on both unweighted and weighted phylogenies. We finally settle almost all questions regarding the nni distance. We show that computing the nni distance is NP-complete (cf. § 2). The proof is quite involved and it uses the lower and upper bounds in [**3, 33, 26**] for sorting on a degree-3 tree by nni operations. The problem is also shown to be NP-complete for *unlabeled* trees, answering another open question in [**3**] (cf. § 3.) We will give an efficient approximation algorithm for computing the nni distance on weighted phylogenies with a performance ratio of $4 \log n + 4$, where $n$ is the number of leaves (cf. § 4). Note that the approximation ratio does not depend on the weights. A special case of this result for unweighted phylogenies was recently reported in [**26**].

Unless otherwise mentioned, all the trees in this paper are trees with all *internal nodes* of degree 3 and with *unique labels* on leaves. We will mention it explicitly if a tree has nonuniquely labeled leaves or unlabeled leaves. Finally, two (weighted) trees are considered equal iff there is an isomorphism between them preserving topology, edge weights, if any, and leaf labels, if any.

**1.1. The nni and Linear-Cost Subtree-Transfer Distances.** The linear-cost subtree-transfer distance was introduced in [**5**]. This distance tries to address biological events such as *recombination* during the course of evolution of molecular sequences of organisms (because of which a single evolutionary tree is no longer sufficient to describe the evolutionary history of the sequences), and gives preferences to those recombinations which are more likely to occur than others. Somewhat surprisingly, although they are studied in parallel for very different reasons, it was demonstrated in [**5**] that the linear-cost subtree-transfer distance is in fact identical to the nni distance for unweighted phylogenies. As a result, all our results in this paper about the nni distance on unweighted phylogenies applies directly to the linear-cost subtree-transfer distance on unweighted phylogenies.

## 2. Computing the nni Distance is NP-complete

THEOREM 1. *Let $T_1$ and $T_2$ be two trees with unique leaf labels and $k$ be an integer. It is NP-complete to decide if $D_{nni}(T_1, T_2) \leq k$.*

Because the proof of Theorem 1 is quite lengthy and complicated, the reader is referred to the appendix for details of the proof. Here we will merely sketch the idea of the proof.

Theorem 1 is proved by a reduction from Exact Cover by 3-Sets (X3C), which is known to be NP-complete [**13**], to our problem. The X3C problem is defined as follows:

INSTANCE: A set $S = \{s_1, \ldots, s_m\}$, where $m = 3q$, and a collection of subsets $C_1, \ldots, C_n$, where $C_i = \{s_{i_1}, s_{i_2}, s_{i_3}\} \subset S$.

QUESTION: Is there an *exact cover* of $S$, that is, are there $q$ disjoint subsets $C_{i_1}, \ldots, C_{i_q}$ such that $\cup_{j=1}^{q} C_{i_j} = S$ ?

Our reduction will construct two trees $T_1$ and $T_2$ with *unique* leaf labels, such that transforming $T_1$ to $T_2$ requires at most $N$ (to be specified later) nni moves iff an exact cover of $S$ exists. The following is an outline of our reduction.

Consider the linear tree shown in Figure 3, where $x_1, \ldots, x_k$ is a sequence of labels. For convenience, such a linear tree will be simply called a *sequence*. Sorting such a sequence means to transform it into another linear tree whose leaves are in a certain desirable order. This can be done by a sequence of nni operations. (During the sorting process, the tree may be nonlinear). The sequences we will construct later consist of small *coding regions* and larger *noncoding regions* that separate the coding regions. Sorting such a sequence will mean sorting each coding region into the ascending order. The $k^2$-sized noncoding regions prevent the merging of adjacent blocks (of a coding region) in an optimal sorting procedure, *i.e.* it will not be beneficial for two blocks with the same corresponding subsequences to be merged and sorted together because it costs at most $ck \log k$ nni moves to sort a block and $k^2$ nni moves to bring a block across a noncoding region.

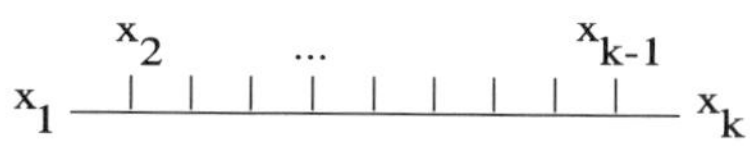


FIGURE 3. A linear tree with $k$ leaves.

To construct the first tree $T_1$, for each $s_i \in S$, we create a sequence $S_i$ of leaves that takes a large number of nni moves to sort. We will make sure that $S_i$ and $S_j$ are "very different" sequences for each pair $i \neq j$, in the sense that we cannot hope to significantly save nni moves by somehow combining the sorting of sequences $S_i$ and $S_j$. Then for each subset $C_i = \{s_{i_1}, s_{i_2}, s_{i_3}\}$, we create three more sequences with the same ordering as the sequences $S_{i_1}, S_{i_2}, S_{i_3}$, respectively, but with distinct labels. Such $n$ groups of sequences for $C_1, \ldots, C_n$, each consisting of three sequences, will be placed "far away" from each other and from the $m$ sequences $S_1, \ldots, S_m$ in tree $T_1$. Tree $T_2$ has the same structure as $T_1$ except that all sequences are sorted and flipped (attached at the other end).

Here is the connection between an exact cover of $S$ and transforming $T_1$ into $T_2$ by nni moves. To transform $T_1$ into $T_2$, all we need is to sort the sequences defined above. If there is an exact cover $C_{i_1}, \ldots, C_{i_q}$ of $S$, we can partition the $m$ sequences $S_1, \ldots, S_m$ into $\frac{m}{3} = q$ groups, according to the cover. For each $C_j$ $(j = i_1, \ldots, i_q)$ in the cover, we send the corresponding group of sequences $S_{j_1}, S_{j_2}, S_{j_3}$ to their counterparts, combine the sorting of the three pairs of sequences with identical ordering, and then transport the three sorted versions of $S_{j_1}, S_{j_2}, S_{j_3}$ back to their original locations in the tree. Thus, instead of sorting six sequences separately, we do three combined sorts involving merging and splitting, plus a round trip transportation of three sequences. Our construction will guarantee that the latter is significantly cheaper. If there is no exact cover of $S$, then either some sequence $S_i$ will be sorted separately or we will have to send at least $q+1$ groups of sequences back and forth. The construction guarantees that both cases will cost significantly more than the previous case.

## 3. Computing the nni Distance Between Unlabeled Trees is NP-Complete

In this section, we prove the NP-completeness of the problem of computing the nni distance between two trees with *unlabeled* or *non-uniquely-labeled* leaves (for the unlabeled case, two trees are identical if and only if they have the same topology). First, we consider the non-uniquely labeled case.

THEOREM 2. *For two given trees $T_1$ and $T_2$ with non-uniquely-labeled leaves and an integer $k$, it is NP-complete to decide if $D_{nni}(T_1, T_2) \leq k$.*

A flawed proof of Theorem 2 was published in [**23**].[1] Although Theorem 2 can be proved by using Theorem 1, here we give a direct and much simpler reduction from the X3C problem (cf. § 2) to this problem.

Assume that we are given the following instance of the X3C problem: A set $S = \{s_1, s_2, \ldots, s_m\}$ with $m = 3q$ and subsets $C_1, C_2, \ldots, C_n$ where $C_i = \{s_{i_1}, s_{i_2}, s_{i_3}\} \subset S$. If $n < q$, $S$ clearly has no exact cover. So we assume $n \geq q$. Let $K > 0$ be an integer to be specified later. We construct two non-uniquely-labeled trees as in Figure 4. There are $n$ *long arms* of length $K$ in $T_1$, $n - q$ long arms of length $K$ and $m$ *short arms* of length $K/3$ in $T_2$. Each of these long (short, respectively) arms is a linear chain with $K$ $(K/3$, respectively) leaves connected to it. These leaves are all identically labeled with a label $x \notin S$. At the end of the $i$-th long arm in $T_1$, we attach three leaves as shown in Figure 4 labeled with the three elements $s_{i_1}, s_{i_2}, s_{i_3}$ in $C_i$, respectively. At the end of the $j$-th short arm in $T_2$, we attach a leaf labeled with $s_j$. The extra $3n - m$ labels (in the multiset $\cup_{i=1}^{n} C_i - S$) not attached to the short arms in $T_2$ are placed between the short and long arms of $T_2$. At the bottom of each long arm in $T_2$, there is no additional labeled leaf.

LEMMA 3. *For any internal edge $e$ within the long arms in $T_1$, the partition of (the multiset of) leaf labels induced by $e$ in $T_1$ is different from the partition of the leaf labels induced by any edge $e'$ in $T_2$.*

PROOF. Every internal edge within a long arm in $T_1$ partitions the leaves of $T_1$ such that one partition contains 3 leaves labeled by the elements of $S$ and an

[1] In [**23**], the author reduced the Partition problem to nni by constructing a tree of $i$ nodes for a number $i$.

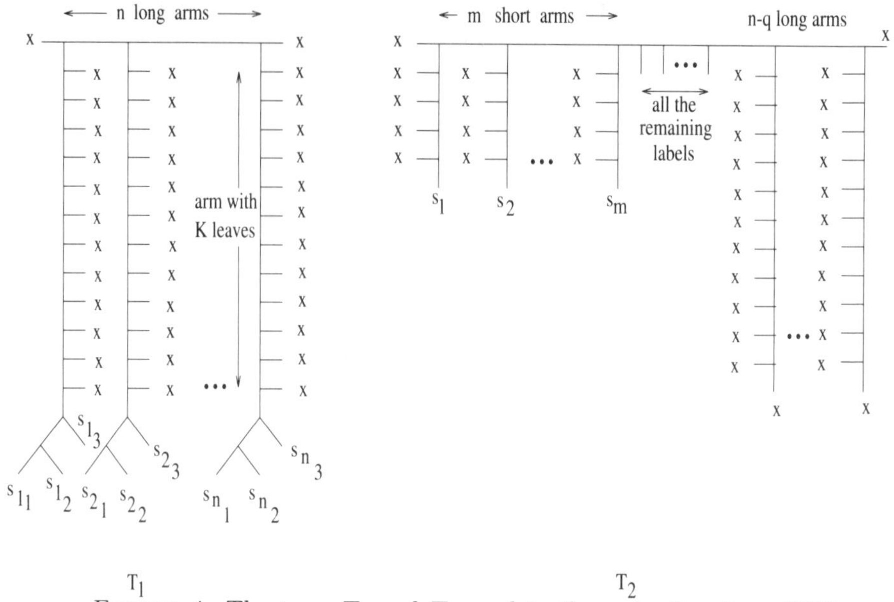


FIGURE 4. The trees $T_1$ and $T_2$ used in the reduction from X3C ($K = 12$).

additional of $p$ leaves ($0 \leq p \leq K$) labeled by $x$. The only internal edges of $T_2$ that partition the leaves with 3 labels from the elements of $S$ has at least $K+1$ leaves labeled by $x$ in that same partition. □

By Lemma 3 we need at least $nK$ nni operations to create all the short arms of $T_2$ from the long arms of $T_1$. We choose $K$ to be a sufficiently large integer (yet polynomial in $m$ and $n$), for example, $K = n^2$. Then the proof of Theorem 2 is completed by proving the following lemma.

LEMMA 4. *For all sufficiently large $m$ and $n$, $D_{nni}(T_1, T_2) \leq nK + o(n^2)$ if and only if there is an exact cover of $S$.*

PROOF. First, assume that there is an exact cover of $S$, say $C_{i_1}, \ldots, C_{i_q}$. Then, $T_1$ can be transformed to $T_2$ by using $nK + o(n^2)$ nni moves in the following manner.

- We transform each of the $q$ long arms in $T_1$ corresponding to $C_{i_1}, \ldots, C_{i_q}$ into three short arms as follows. Consider the long arm $L$ corresponding to $C_{i_j} = \{s_{l_1}, s_{l_2}, s_{l_3}\}$. (See Figure 5 for an illustration). Leave the leaf node with label $s_{l_3}$ at the bottom of $L$, and move the two leaf nodes with labels $s_{l_1}$ and $s_{l_2}$ up $L$ by a distance $K/3$. Then leave the leaf node with label $s_{l_1}$ there, and move the remaining leaf node with label $s_{l_2}$ together with the linear subtree of $L$ of the last $K/3 + 1$ nodes up by a distance $K/3$. Now, leave the leaf node with label $s_{l_2}$ there, and move two linear subtrees up. This in total needs $K + 4$ nni moves for each long arm.
- For each long arm corresponding to a subset $C_l$ not in the exact cover, we simply move the three leaf nodes at the bottom up (see Figure 6). This needs $K + 3$ nni moves.

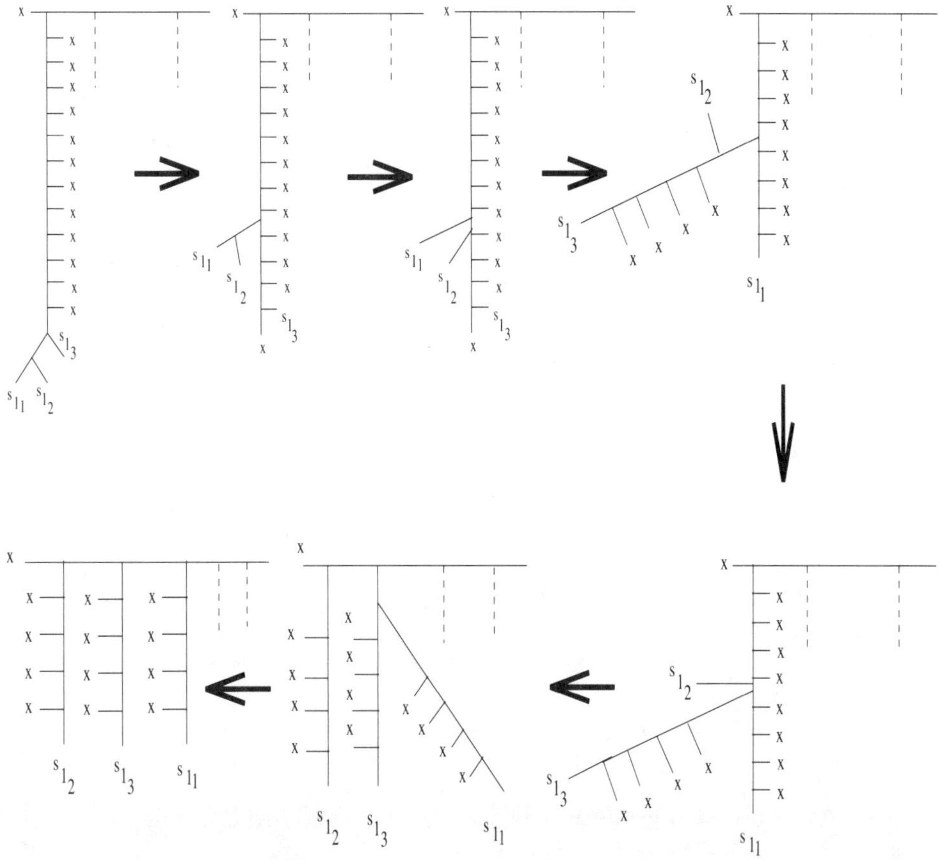


FIGURE 5. Transforming a long arm into three short arms.

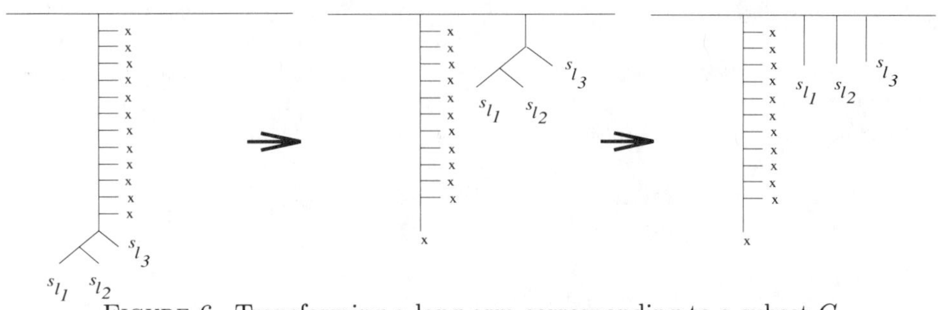


FIGURE 6. Transforming a long arm corresponding to a subset $C_l$ not in the exact cover of $S$.

- Now, we have created the short and long arms of $T_2$, but not necessarily in the order they appear in $T_2$. Since $n \geq q = \frac{m}{3}$, it is trivial to use at most $o(n^2)$ nni operations[2] to shuffle the short and the long arms to finally obtain $T_2$.

[2]Obviously, this is a trivial upper bound which can be further improved. However, such improvements are not necessary for the correctness of this proof.

The total number of nni moves we use is:

$$(K+4)\frac{m}{3} + (K+3)(n - \frac{m}{3}) + o(n^2) \leq nK + o(n^2),$$

as claimed.

Conversely, assume that there is no exact cover of $S$. Then, to convert $T_1$ to $T_2$, by Lemma 3, the $nK$ nni moves are still necessary to move all leaves with labels in $S$ up and create the short and long arms in $T_2$. However, since we do not have an exact cover, either one leaf with label in $S$ must move down at least a short arm, or a short arm of length $K/3$ must merge with another short arm. Both costs at least $K/3$ extra nni moves. Thus in total, we must use at least $nK + K/3 > nK + o(n^2)$ nni moves (by our choice of $K$ for sufficiently large values of $n$). This completes the proof. □

Next, we consider the case when the leaves are unlabeled.

THEOREM 5. *For any two given trees $T_1$ and $T_2$ with unlabeled leaves and an integer $k$, it is NP-complete to decide if $D_{nni}(T_1, T_2) \leq k$.*

PROOF. Again, we reduce the X3C problem to this problem. In Figure 4, the trees $T_1$ and $T_2$ use $m+1$ different labels: $s_1, \ldots, s_m$ and $x$. We denote $x$ by $s_{m+1}$. For each leaf node with label $s_i$ ($1 \leq i \leq m+1$) in $T_1$ and $T_2$, we replace the node by a linear tree $L_i$ of length $iK^2$. Let $T_1'$ and $T_2'$ be the resulting two trees (with unlabeled leaves). Since these linear trees $L_i$'s are non-isomorphic and have distance of at least $K^2$ to each other, in order to transform $T_1'$ to $T_2'$, they have to be moved entirely to satisfy the bounds in Lemma 4 if there is an exact cover. Thus, the claim in Lemma 4 still holds, which implies Theorem 5. □

## 4. Approximating the nni Distance on Weighted Phylogenies

In this section, we present an $O(n^2)$ time approximation algorithm with performance ratio $4 \log n + 4$ for computing the nni distance on weighted phylogenies. As mentioned in § 1, many phylogeny reconstruction methods produce weighted phylogenies. Hence the weighted nni distance problem is also very important in computational molecular biology. This algorithm generalizes the approximation algorithm in **[26]** for unweighted phylogenies, but is considerably more complicated.

Obviously, the nni operations can be performed only on internal edges and they do not change the weight of any edge. Given two weighted trees $T_1$ and $T_2$, for feasibility of transformation between $T_1$ and $T_2$ by using nni moves, we require that the following *feasibility conditions* are satisfied:

**(i):** For each leaf label $a$, the weight of the edge in $T_1$ incident to $a$ is the same as the weight of the edge in $T_2$ incident to $a$.

**(ii):** The multisets of the weights of internal edges of $T_1$ and $T_2$ are identical.

If $T_1$ and $T_2$ do not satisfy the above two conditions, clearly $T_1$ cannot be transformed to $T_2$ by using nni operations.

DEFINITION 1. Let $T_1$ and $T_2$ be two weighted trees with (not necessarily uniquely) labeled leaves. An internal edge $e_1$ of $T_1$ and an internal edge $e_2$ of $T_2$ are a *good edge pair* iff the following hold:

1. $w(e_1) = w(e_2)$.
2. The partition (of the multiset) of edge weights induced by $e_1$ in $T_1$ is the same as the partition of edge weights induced by $e_2$ in $T_2$.

3. The partition (of the multiset) of leaf labels induced by $e_1$ in $T_1$ is the same as the partition of leaf labels induced by $e_2$ in $T_2$.

Intuitively, if $e_1$ and $e_2$ form a good edge pair, then in order to transform $T_1$ to $T_2$, it is not necessary to perform an nni operation on $e_1$. An edge $e_1 \in T_1$ is *bad* if there is no edge $e_2$ in $T_2$ such that $e_1$ and $e_2$ form a good edge pair. Definition 1 applies to unweighted trees by letting all edges to have weight 1.

THEOREM 6. *Let $T_1$ and $T_2$ be two weighted phylogenies each with n leaves. Then, $D_{nni}(T_1, T_2)$ can be approximated to within a factor of $4(1 + \log n)$ in $O(n^2)$ time.*

In the rest of this section, we prove Theorem 6. The basic idea of the algorithm is as follows. We first identify "bad" components in the tree that need a lot of nni moves in the transformation process. Then, for each bad component, we put things in correct order by first converting them into balanced shapes. But notice that we cannot afford to perform nni operations many times on the edges with heavy weights. Furthermore, not only the leaf nodes need to be moved to the right places, so do the weighted edges. The main difficulty of our algorithm is the careful coordination of the transformations so that at most $O(\log n)$ nni operations are performed on each heavy edge.

Note that given an adjacency-list representation of a tree, it takes $O(1)$ time to update the tree after a single nni operation. Since the multisets of internal edge weights of $T_1$ and $T_2$ are the same, for simplicity, we use $\{e_1, e_2, \ldots, e_{n-3}\}$ to denote the set of internal edges of both $T_1$ and $T_2$. We renumber the edges of both $T_1$ and $T_2$, if necessary, in $O(n \log n)$ time such that $w(e_1) \leq w(e_2) \leq \ldots \leq w(e_{n-3})$. Let $W = \sum_{i=1}^{n-3} w(e_i)$.

The following lemma provides a lower bound on $D_{nni}(T_1, T_2)$, which is needed to establish the performance ratio of our approximation algorithm. It holds for either unweighted or weighted trees.

LEMMA 7. *If $T_1$ and $T_2$ have no good edge pairs, then $D_{nni}(T_1, T_2) \geq W$.*

PROOF. For each internal edge $e_i \in T_1$, the partition of either the leaf labels or the edge weights induced by $e_i$ is different from that induced by any edge in $T_2$. Hence, in order to transform $T_1$ into $T_2$, at least one nni operation must be performed on $e_i$ with cost $w(e_i)$. So the total cost of transforming $T_1$ into $T_2$ is at least $W$. □

We are ready to present our algorithm. First we consider the special case where $T_1$ and $T_2$ have no good edge pairs. By Lemma 7 it suffices to describe how to transform $T_1$ to $T_2$ with total cost at most $(4 \log n + 4)W$. The algorithm for this case consists of three steps.

*Step 1*: Pick an arbitrary leaf node $r$ and transform $T_1$ into a balanced binary tree $T_1'$ of height $\lceil \log n \rceil$ as shown in Figure 7, where the internal edges $e_i$ $(1 \leq i \leq n-3)$ are positioned as follows: at the $i$th level $(i \geq 1)$, $e_{2^i-1+j}$ $(0 \leq j < 2^i)$ is the $j$th edge from the left. It is easy to see that, on any path from $r$ to a leaf, the weight of the internal edges are non-decreasing. This fact will be needed in Step 3 of our algorithm. Step 1 is carried out in three phases.

*Phase 1.1*: Transform $T_1$ to a linear tree $L$ as shown in Figure 8, where the internal edges $e_1, \ldots, e_{n-3}$ appear in some arbitrary order form left to right as in Figure 8. This can be done as follows: Treat $T_1$ as an rooted and ordered tree with

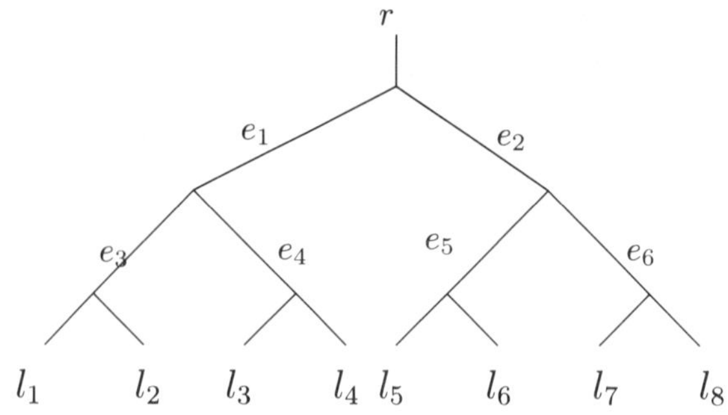


FIGURE 7. The balanced tree $T_1'$ when $n = 9$.

root $r$. Each edge of $T_1$ will be either a *left* or a *right* edge. The edge from $r$ to its child is drawn as a left edge. Consider any internal node $u$ with children $v_1$ and $v_2$. If both $v_1$ and $v_2$ are internal nodes, or both are leaf nodes, we may chose either the edge $(u, v_1)$ or the edge $(u, v_2)$ as the left edge. If only one child (say $v_1$) is an internal node, we chose $(u, v_1)$ as the left edge. The *left path* of $T_1$ is the path $P$ from $r$ to a leaf node using only left edges. If $P$ contains all internal edges, then $T_1$ is already a linear tree and we are done. Otherwise, let $e$ be an right internal edge with one end node on $P$. Perform an nni operation on $e$. Re-arranging the left and the right edges according to the above description, we obtain a new tree whose left path contains one more edge than the left path $P$ of $T_1$. Repeat this process until $T_1$ is transformed into a linear tree. Clearly, at most one nni operation is performed on each internal edge of $T_1$ during the transformation process. Thus Phase 1.1 costs at most $W$ and can be completed in $O(n)$ time.

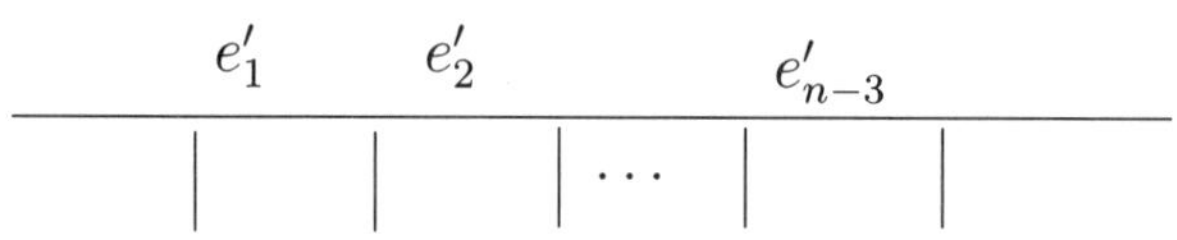


FIGURE 8. The linear tree $L$.

*Phase 1.2*: Similar to Phase 1.1, we transform $T_1'$ to some linear tree $L'$ in $O(n)$ time with total cost at most $W$. Let the internal edges appear in $L'$ from left to right as

$$e_1'', e_2'', \cdots, e_{n-3}''. \tag{1}$$

*Phase 1.3*: We use an analogue of merge sort to transform the linear tree $L$ to the linear tree $L'$. The transformation is the same as in [**26**], but focuses on the internal edges instead. This transformation costs $W \log n$ and can be performed in $O(n \log n)$ time.

To transform $T_1$ to $T_1'$, we perform the nni operations in Phase 1.1, followed by the nni operations in Phase 1.3, followed by the inverse of the nni operations in Phase 1.2. Thus Step 1 can be completed in $O(n \log n)$ time with total cost at most $(2 + \log n)W$.

*Step 2*: Transform $T_2$ to a balanced binary tree $T_2'$. The position of the internal edges of $T_2'$ are identical to that of $T_1'$. Similar to Step 1, this can be done in $O(n \log n)$ time with total cost at most $(2 + \log n)W$.

*Step 3*: Transform $T_1'$ to $T_2'$. Since both trees have identical internal structure, we only need to move the leaves of $T_1'$ to their corresponding positions in $T_2'$. Let $r = l_0, l_1. \dots, l_{n-1}$ denote the leaf nodes in $T_1'$ in counter-clockwise order starting at the root leaf $r$. Let $l_{i'}$ be the destination in of the leaf $l_i$ $(1 \leq i \leq n-1)$ in $T_2'$. Denote the permutation mapping $i$ to $i'$ by $\pi$.

We move the leaves in $T_1'$ according to $\pi$. Write $\pi$ as a product of disjoint cycles, and process each cycle $C = (i_1, i_2, \dots, i_k)$ in turn as follows. The lowest internal edge above the leaf $l_{i_j}$ will be called the *internal edge adjacent to* $l_{i_j}$. Without loss of generality, assume $l_{i_1}$ have the lightest adjacent internal edge weight among all $l_{i_j}$ $(1 \leq j \leq k)$. Now move the leaf $l_{i_1}$ to the leaf $l_{\pi(i_1)} = l_{i_2}$ by a sequence of nni operations. Then, move $l_{i_2}$ to $l_{i_3}$ and so on. The last nni operation that swaps $l_{i_1}$ into place also starts $l_{i_2}$ on its way up and we continue moving it to $l_{i_3}$. Finally, leaf $l_{i_k}$ is moved to take the place vacated by $l_{i_1}$. This completes the processing of the cycle $C$. It is easy to see that this process restores the original topology of internal edges. Figure 9 illustrates this process for a cycle $(1, 8, 3)$.

Since the height of $T_1'$ is $\lceil \log n \rceil$, it is easy to see that Step 3 takes at most $O(n \log n)$ time. Next, we analyze the total cost of Step 3. For simplicity, we assume that $n = 2^m + 1$ (see Figure 7). Partition the set of the internal edges of $T_1'$ into $E_1 \cup \dots \cup E_{m-1}$, where $E_i$ consists of the internal edges at level $i$. Let $W_i$ be the total weight of the edges in $E_i$. For any $e \in E_i$, there are $2^{m-i}$ leaves below $e$. Thus there are at most $2^{m-i}$ nni operations that move a leaf "down" $e$ and at most $2^{m-i}$ nni operations that move a leaf "up" $e$. Furthermore, consider the processing of a cycle $C = (i_1, i_2, \dots, i_k)$. Let $e_{i_j}$ $(1 \leq j \leq k)$ be the internal edge in $E_{m-1}$ that is above the leaf $l_{i_j}$. Let $w(C)$ denote the total weight of the edges $e_{i_1}, \dots, e_{i_k}$. During the processing of $C$, the up and down nni operations coincide for each $e_{i_j}$ $(2 \leq j \leq k)$. Thus the cost contribution of the edges in $E_{m-1}$ for processing $C$ is at most: $2w(e_{i_1}) + \sum_{j=2}^{k} w(e_{i_j})$ which, by our choice of cycle start $i_1$, is at most $1.5w(C)$. Summing up over all cycles $C$ in $\pi$, the total cost $D$ for Step 3 is at most:

$$D \leq 3W_{m-1} + \sum_{i=1}^{m-2} 2^{m-i+1} W_i = \sum_{i=1}^{m-1} 2^{m-i+1} W_i - W_{m-1}.$$

In the linear combination $\omega = \sum_{i=1}^{m-1} 2^{m-i+1} W_i$, the coefficient of $W_i$ is twice that of $W_{i+1}$. By our choice of the ordering of the internal edges, we have $W_i \leq W_{i+1}/2$. It follows that $\omega$ is maximized when $W_i = W_{i+1}/2$, i.e. all edges have uniform weight. Together with the only other constraint $\Sigma_{i=1}^{m-1} W_i = W$ and noting that there are $2^m - 2$ internal edges in all, this gives $W_i = 2^i * W/(2^m - 2)$ for the worst case cost. Assuming for convenience that $m \geq 5$, so that $2^{m-1}/2 \geq 2(m-1)$, we have:

$$(2) \quad D \leq \sum_{i=1}^{m-1} 2^{m-i+1} W_i - W_{m-1} = \frac{(m-1)2^{m+1} - 2^{m-1}}{2^m - 2} W \leq 2(m-1)W$$

Thus the total cost of Step 3 is at most $2(\log n)W$.

To transform $T_1$ to $T_2$, we first perform the nni operations in Step 1 (transforming $T_1$ to $T_1'$), followed by the nni operations in Step 3 (transforming $T_1'$ to

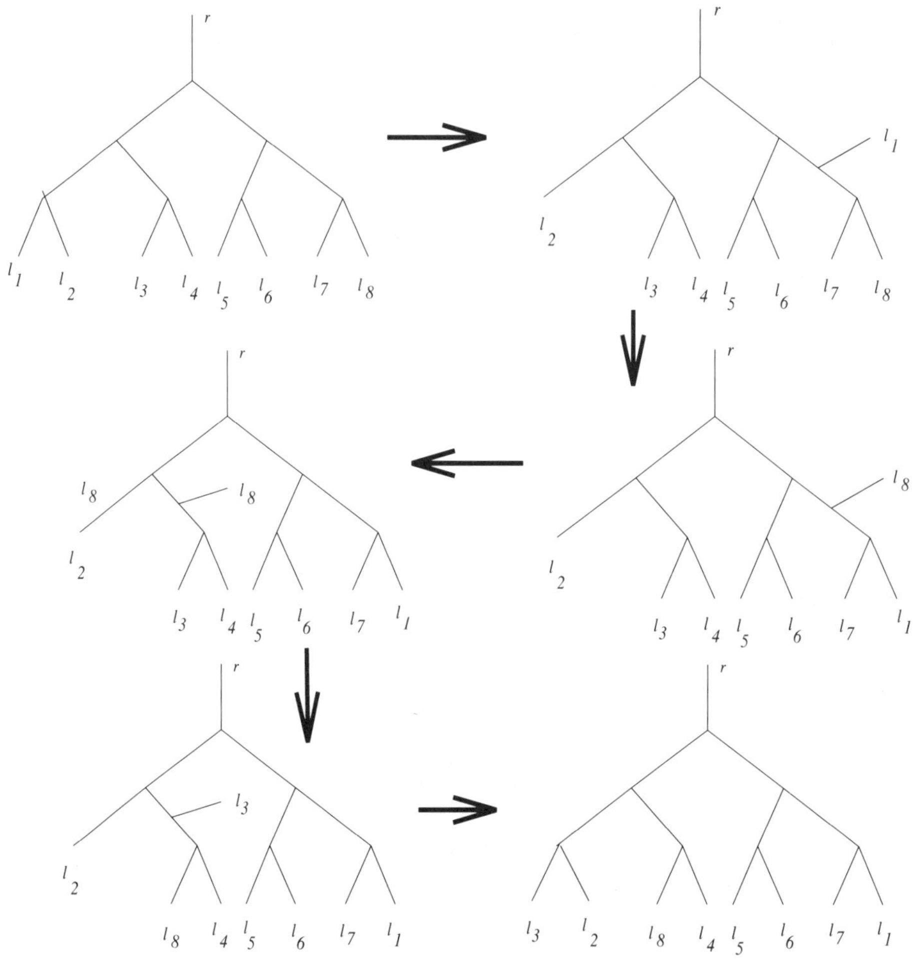

FIGURE 9. The nni operations for processing the cycle $(1, 8, 3)$.

$T_2'$), followed by the inverse of the nni operations in Step 2 (transforming $T_2'$ to $T_2$). The complete algorithm can be done in $O(n \log n)$ time with total cost at most $4(1 + \log n)W$.

Next we consider the general case when $T_1$ and $T_2$ may have some good edge pairs. We first need to identify the set $E''$ of edges in $T_1$ that form good edge pairs with edges in $T_2$. The proof of the following lemma can be found in [**7**].

LEMMA 8. [**7**] *Let $T_1$ and $T_2$ be two trees, each with $n$ leaves. Then, the set of edges of $T_1$ which partition the leaf labels similarly as some edge of $T_2$ can be found in $O(n)$ time.*

Using the above lemma, it is trivial to find the set $E''$ in $O(n^2)$ time: First, we find the set of all edges $E''' \supseteq E''$ of $T_1$ that partition the leaf labels similarly as some edge of $T_2$ in $O(n)$ time using Lemma 8. It is trivial to identify the edges in $E''$ that also satisfy the other conditions of Definition 1, in a total of $O(n^2)$ time.

Let $E'$ be the set of internal edges of $T_1$ not in $E''$. Similar to Lemma 7, we can show that at least one nni operation on each internal edge $e \in E'$ is needed to transform $T_1$ to $T_2$. Thus:

$$D_{nni}(T_1, T_2) \geq W' = \sum_{e \in E'} w(e) \tag{3}$$

The edges in $E'$ induce in $T_1$ a subgraph consisting of one or more connected components each of which is a subtree of $T_1$. These connected components can easily be found in $O(n)$ time. To transform $T_1$ to $T_2$, we perform the approximation algorithm described above on each such component. The total cost is bounded by $4(1 + \log n)W'$. The algorithm takes $O(n^2)$ time. This completes the proof of Theorem 6.

## 5. Conclusion and Open Problems

The results reported in this paper have been obtained as a part of our project of building a software package for comparing phylogenetic trees. Several open questions still remain:

- Can we approximate the nni distance with a better ratio (on weighted or unweighted phylogenies)? It seems that to obtain a ratio of $o(\log n)$, we have to be able to prove nontrivial lower bounds for sorting sequences on trees with nni moves.
- The nni operation is similar to and slightly more powerful than the *rotation operation* discussed in [**3, 32**]. Is it NP-complete to compute the rotation distance? Can we approximate the rotation distance better than the trivial ratio 2? This question turns out to be subtler than it appears to be. Partial results in this direction can be found in [**27**].

**Acknowledgments:** We wish to thank J. Felsenstein for explaining to us the biological motivation for comparing weighted phylogenies.

## References

[1] D. Barry and J.A. Hartigan, Statistical analysis of hominoid molecular evolution, *Stat. Sci.*, 2(1987), 191-210.

[2] R. P. Boland, E. K. Brown and W. H. E. Day, Approximating minimum-length-sequence metrics: a cautionary note, *Math. Soc. Sci.*, 4(1983), 261–270.

[3] K. Culik II and D. Wood, A note on some tree similarity measures, *Inform. Proc. Let.*, 15(1982), 39–42.

[4] B. DasGupta, X. He, T. Jiang, M. Li, J. Tromp and L. Zhang, On distances between phylogenetic trees, in *Proc. 8th ACM-SIAM Symposium on Discrete Algorithms*, 1997, pp. 427-436.

[5] B. DasGupta, X. He, T. Jiang, M. Li, and J. Tromp, On the linear-cost subtree-transfer distance, to appear in the special issue in *Algorithmica* on computational biology, 1998.

[6] W. H. E. Day, Properties of the nearest neighbor interchange metric for trees of small size, *Journal of Theoretical Biology*, 101(1983), 275–288.

[7] W. H. E. Day, Optimal algorithms for comparing trees with labeled leaves, *J. Classification*, 2(1985), 7-28.

[8] A.W.F. Edwards and L.L. Cavalli-Sforza, The reconstruction of evolution, *Ann. Hum. Genet.*, 27(1964), 105. (Also in *Heredity* 18, 553.)

[9] J. Felsenstein, Evolutionary trees for DNA sequences: a maximum likelihood approach. *J. Mol. Evol.*, 17(1981), 368-376.

[10] J. Felsenstein, personal communication, 1996.

[11] W.M. Fitch, Toward defining the course of evolution: minimum change for a specified tree topology, *Syst. Zool.*, 20(1971), 406-416.

[12] W.M. Fitch and E. Margoliash, Construction of phylogenetic trees, *Science*, 155(1967), 279-284.

[13] M. R. Garey and D. S. Johnson, *Computers and Intractability: A Guide to the Theory of NP-Completeness,* W. H. Freeman, 1979.

[14] Arndt von Haseler and Gary A. Churchill, Network models for sequence evolution, *J. Mol. Evol.*, 37(1993), 77-85.

[15] J. Hein, Reconstructing evolution of sequences subject to recombination using parsimony, *Math. Biosci.*, 98(1990), 185–200.

[16] J. Hein, A heuristic method to reconstruct the history of sequences subject to recombination, *J. Mol. Evol.*, 36(1993), 396–405.

[17] J. Hein, personal email communication, 1996.

[18] J. Hein, T. Jiang, L. Wang, and K. Zhang, On the complexity of comparing evolutionary trees, *Discrete Applied Mathematics* 71, 153-169, 1996.

[19] J. P. Jarvis, J. K. Luedeman and D. R. Shier, Counterexamples in measuring the distance between binary trees, *Mathematical Social Sciences*, 4(1983), 271–274.

[20] J. P. Jarvis, J. K. Luedeman and D. R. Shier, Comments on computing the similarity of binary trees, *Journal of Theoretical Biology* 100(1983), 427–433.

[21] S. Khuller, Open Problems: 10, *SIGACT News*, 24:4(Dec., 1994), p.46.

[22] V. King and T. Warnow, On measuring the nni distance between two evolutionary trees, *DIMACS Mini Workshop on Combinatorial Structures in Molecular Biology*, Rutgers University, Nov 4, 1994.

[23] M. Křivánek, Computing the nearest neighbor interchange metric for unlabeled binary trees is NP-complete, *Journal of Classification* 3(1986), 55–60.

[24] M. Kuhner and J. Felsenstein, A simulation comparison of phylogeny algorithms under equal and unequal evolutionary rates. *Mol. Biol. Evol.* 11(3), 1994, 459–468.

[25] W.J. Le Quesne, The uniquely evolved character concept and its cladistic application, *Syst. Zool.*, 23(1974), 513-517.

[26] M. Li, J. Tromp, and L. Zhang, Some notes on the nearest neighbor interchance distance, *Journal of Theoretical Biology*, 182(1996), 463-467.

[27] M. Li and L. Zhang, Better approximation of diagonal-flip transformation and rotation transformation. *COCOON98*, Aug. 12-14, 1998, Taipei, Taiwan.

[28] G. W. Moore, M. Goodman and J. Barnabas, An iterative approach from the standpoint of the additive hypothesis to the dendrogram problem posed by molecular data sets, *Journal of Theoretical Biology* 38(1973), 423–457.

[29] D. F. Robinson, Comparison of labeled trees with valency three, *Journal of Combinatorial Theory, Series B,* 11(1971), 105–119.

[30] N. Saitou and M. Nei, The neighbor-joining method: a new method for reconstructing phylogenetic trees, *Mol. Biol. Evol.*, 4(1987), 406-425.

[31] D. Sankoff, Minimal mutation trees of sequences, *SIAM J. Appl. Math.*, 28(1975) 35-42.

[32] D. Sleator, R. Tarjan and W. Thurston, Rotation distance, triangulations, and hyperbolic geometry, *J. Amer. Math. Soc.*, 1(1988), 647-681.

[33] D. Sleator, R. Tarjan and W. Thurston, Short encodings of evolving structures, *SIAM J. Discr. Math.*, 5(1992), 428–450.

[34] M. S. Waterman and T. F. Smith, On the similarity of dendrograms, *Journal of Theoretical Biology*, 73(1978), 789–800.

## APPENDIX

**Proof of Theorem 1:** Using the discussion and notations in § 2, we now continue to give the detailed proof of Theorem 1. Apparently many difficult questions have to be answered: How can we find these $m$ sequences $S_1, \ldots, S_m$ that are *hard* to sort by nni moves? How do we ensure that sorting one such sequence will never help to sort others? How can we ensure that it is most beneficial to bring the sequences $S_{j_1}, S_{j_2}, S_{j_3}$ corresponding to elements in a subset $C_j$ to their counterparts to get sorted, and not the other way around?

We begin with the construction of the sequences $S_1, \ldots, S_m$. Recall that each such sequence is actually a linear tree, as in Figure 3. Intuitively, it would be a good idea to take a long and difficult-to-sort sequence and break it into $m$ pieces of equal length. But because the upper bound in [**3, 26**] and the lower bound in [**33**] (see [**26**] for the calculation) do not match, this simple idea does not work. The reason is that it is impossible to guarantee that sorting one of these pieces will not help to sort the other pieces (by merging, sorting together, and then splitting). So we first have to find sequences that are hard to sort and do not help each other in sorting.

LEMMA 9. *For any constant $\epsilon > 0$, there exists a constant $c > \epsilon$ and infinitely many $k$ for which there are two sequences $x$ and $y$ of length $k$ such that (i) each of them takes at least $(c-\epsilon)k\log k$ nni moves to sort, and (ii) each of them takes at most $ck\log k$ nni moves to sort, and (iii) it takes at least $(c-\epsilon)(2k)\log(2k)$ nni moves to sort both of them together, i.e. the sequence $xy$.*

PROOF. From the results in [**3, 26, 33**], we know that for each $k$, there exists a sequence of $k$ leaves such that sorting the sequence takes at most $k\log k + O(k)$ nni moves and at least $\frac{1}{4}k\log k - O(k)$ nni moves. Let us define $c_k$, for any $k$, as the maximum number of nni moves to sort any sequence of length $k$, divided by $k\log k$. Since $\frac{1}{4} - o(1) \le c_k \le 1 + o(1)$, infinitely many $k$ satisfy

$$c_{2k} \ge c_k - \frac{\epsilon}{3}. \tag{4}$$

Since any length $2k$ sequence can be sorted by first sorting both length $k$ halves and then merging them using less than $2k$ nni moves, we have $c_{2k}2k\log(2k) < 2c_k k\log k + 2k$ hence $c_{2k} \le c_k + o(1)$. Taking $xy$ to be a hardest sequence of length $2k$ for large enough $k$ satisfying the inequality (4), and taking $c = c_k$, necessarily satisfies conditions (ii) and (iii). Without loss of generality, let $w_x \le w_y$ be the costs of sorting $x$ and $y$. Then the above sorting method for $xy$ shows that $c_{2k}2k\log(2k) \le w_x + w_y + 2k$ hence, together with the fact that $w_y \le ck\log k$, we have:

$$\begin{aligned} w_y \ge w_x &\ge (c - \frac{\epsilon}{3})2k(1+\log k) - 2k - ck\log k \ge (c - \frac{2\epsilon}{3})k\log k - 2k(1 + \frac{\epsilon}{3} - c) \\ &\ge (c-\epsilon)k\log k \end{aligned}$$

proving (i). Furthermore, infinitely many of these $c_k$ must be arbitrarily close to each other, giving a single $c$ to satisfy the lemma. □

Let $\epsilon = 1/8$, $k$ a sufficiently large integer satisfying Lemma 9 and $c, x, y$ the corresponding constant and sequences. Next we use $x$ and $y$, each of length $k$, to construct $m$ long sequences $S_1, \ldots, S_m$. Choose $m$ binary strings $\alpha_1, \ldots, \alpha_m$ in $\{0,1\}^{\lceil \log m \rceil}$. The sequence $S_i$ $(1 \le i \le m)$ is obtained from the string $\alpha_i$ by performing the following: Replace each letter 0 with the sequence $x^{m^3}$ and each letter 1 with the sequence $y^{m^3}$. Give each occurrence of $x$ and $y$ unique labels. Insert between every pair of adjacent $x/y$ blocks a *delimiter* sequence of length $k^2$ with unique labels. This results in sequences $S_1, \ldots, S_m$, all with distinct labels. Each pair of them differ by at least $m^3$ $x$ or $y$ blocks at some position. The $m$ sequences will have specific orientations in the tree; let's refer to one end as *head* and the other end as *tail*.

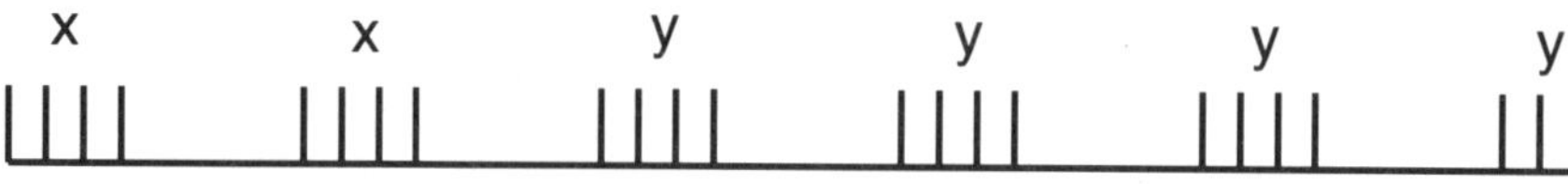


FIGURE 10. The structure of the string $S_i$ corresponding to binary string 011.

Figure 10 shows the structure of a single sequence using binary string 011. To make the picture clear, we pretend $m^3$ is only 2. Assume both $x$ and $y$ blocks have length $k = 4$. The gaps between blocks represent the noncoding regions whose leaves are not shown explicitly. Recall that sorting this sequence entails sorting all the blocks plus moving the point of attachment (connecting point to the tree) from the left to the right of the sequence. The latter requires $k^2(m^3 \log m - 1)$ moves to cross the noncoding regions. To this we need to add for each block the cost of sorting it and having the attachment move through it. Let $q_i$ denote the sum of the cost of sorting and passing through each block in the sequence.

We are now ready to do the reduction. From the set $S = \{s_1, \ldots, s_m\}$ and the subsets $C_1, C_2, \ldots, C_n$, we construct the tree $T_1$ as follows. (The construction of $T_2$ will be described later). For each element $s_i \in S$, $T_1$ has a sequence $S_i$ as defined above. For each subset $C_i = \{s_{i_1}, s_{i_2}, s_{i_3}\}$, we create three sequences $S_{i,i_1}, S_{i,i_2}, S_{i,i_3}$, with the same ordering as $S_{i_1}, S_{i_2}, S_{i_3}$, respectively, but with different and unique labels. Notice that we are not allowed to repeat labels.

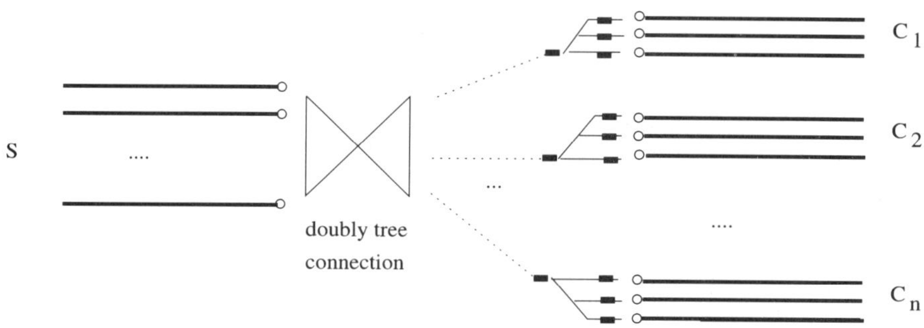


FIGURE 11. The structure of the tree $T_1$

Figure 11 outlines the structure of the tree $T_1$. Here a thick solid line represents a sequence $S_i$ or $S_{i,j}$ with the circled end as head; a dotted line represents a *toll* sequence of $m^2$ uniquely labeled leaves; and a small black rectangle represents a **one-way circuit** as illustrated in Figure 13(i). The heads of the $m$ sequences at the left of Figure 11 are connected by a (double) tree-type connection of depth $\log m + \log n$ to the $n$ toll sequences, each leading to the *entrance* of a one-way circuit. The *exit* of each such one-way circuit is connected to the entrances of three one-way circuits leading finally to the three sequences corresponding to some subset $C_i$.

As mentioned before, a sequence $S_i$ and a counterpart $S_i'$ with identical ordering will be brought together to be sorted. We now consider the cost of combining the sorting process of $S_i$ and $S_i'$. At the beginning of this process, the attachments of $S_i$ and $S_i'$ (i.e. the edge linking it to the main structure of $T_1$) are at the left end of $S_i$ and $S_i'$, respectively (see Figure 12). During the sorting process, the attachments move through $S_i$ and $S_i'$ from left to right and their blocks are sorted on the way. At the end of the process, the attachments are at the right end of $S_i$ and $S'_i$, with all their blocks sorted. In Figure 12, the top (bottom, respectively) long horizontal line represents the sequence $S_i$ ($S'_i$, respectively). The bridge between the two sequences represents the attachments to the main structure (which is not shown). Consider the processing of a block $B$ in $S_i$ and its corresponding block $B'$ in $S'_i$. The attachments move from left to right. Each leaf in $B'$ moves four steps across the attachments to join its partner in $B$, followed by one move to advance the attachments. (In Figure 12, one leaf of $B'$ has joined its partner in $B$ and the attachment has passed one leaf of $B$). After $5k$ moves, all leaves of $B'$ have joined their partners in $B$ and the attachments are at the right of $B$. We spend another $k$ moves to get attachment of $S_i$ back to the left of $B$. Now we can proceed as in the single sequence case to sort the combined blocks $B$ and $B'$. (This will move the attachments to the right of the block $B$ again). Finally, we move the attachment of $S$ from right to left again in $k$ moves (to get ready for splitting process), and spend another $5k$ moves to split $B$ and $B'$. At the end, both $B$ and $B'$ are sorted and the attachments end up on the right of both blocks. The total cost for processing $B$ and $B'$ is thus $12k$ plus the cost of sorting a single block $B$.

Compared to sorting $S_i$ and $S'_i$ separately, because we saved $q_i$ for sorting (and passing through) all coding blocks of a sequence, but spent extra $12k$ moves on each coding block, we thus save

$$q_i - 12km^3 \log m \tag{5}$$

nni moves.

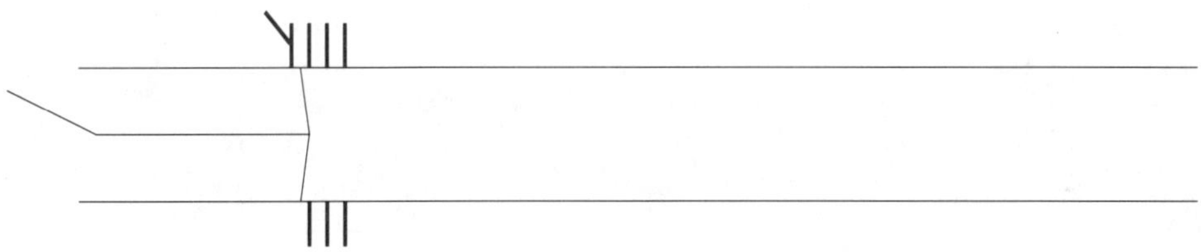

FIGURE 12. Merging two sequences.

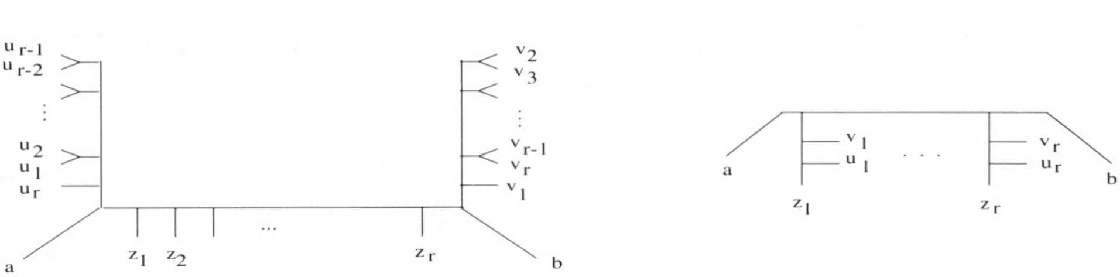


FIGURE 13. One-way circuit.

A one-way circuit is shown in Figure 13 (i) (where $r$ is an integer to be determined later). The leaf labeled by $a$ ($b$, respectively) is the *entrance* (*exit*, respectively) of the one-way circuit. The counterpart of the one-way circuit in $T_2$ is as shown in Figure 13 (ii). It is designed for the purpose of giving "free rides" to a subtree moving first from the entrance to the exit and then later from the exit back to the entrance, while transforming the one-way circuit (i) into its counterpart (ii).

On the other hand, it imposes a large extra cost for subtrees first moving from the exit to the entrance and then back to the exit. This can be seen as follows.

In any optimal transformation of circuit (i) to (ii), the leaves marked by $u$'s are paired up with the leaves marked by $z$'s first and then the leaves marked by $v$'s are paired with the $u$-$z$ pairs. This requires $u_r$ and $v_1$ to move up and out of the way. The pairing of the $u$'s essentially provides a shortcut for $u_r$ to reach $z_r$ in half as many steps, and similarly for $v_1$.

A precise breakdown of the cost is as follows: $(r-3)/2$ steps to move $u_r$ up, then $\frac{r-1}{2}$ times 6 steps to move each $u$ pair down between the proper $z$'s and pair them up, and one final step to pair $u_r$. The exact same number of steps is needed for the symmetric pairing of $v$'s. Hence in total we need (assuming $r$ is odd)

$$2(\frac{r-3}{2}+6\frac{r-1}{2}+1)=r-3+6(r-1)+2=7r-7$$

nni moves. Note that a subtree situated at 'a' can initially pair up with $u_r$ in 2 steps and move together with it, spending 3 more steps to pop off just before $u_r$ pairs with $z_r$, to end up at 'b'. It can later spend another 5 steps to move together with $v_1$ ending up back at 'a'. A subtree going first from 'b' to 'a' and then back to 'b' could only be done 'for free' by pairing with $v_1$ first and with $u_r$ later, since these are the only leaves to move away from 'b' and 'a' respectively in an optimal transformation. But for $v_1$ to reach 'a' with minimum cost requires collapsing all the $v$'s which imposes an extra cost on pairing $u$'s with $z$'s later. The least penalty for moving from 'b' to 'a' back to 'b' is thus for $v_1$ not to take the shortcut which costs an extra $\frac{r}{2}$ steps.

We will choose $r$ so large (*i.e.* $r=m^4$) that it is not worthwhile to move any sequence $S_{i,j}$, corresponding to some $C_i$, to the left through the one-way circuits to sort and then move it back to its original location in $T_1$.

In the following *sorting* a sequence $S_i$ or $S_{i,j}$ means to have each of its $x/y$ blocks sorted and then the whole sequence *flipped.* The tree $T_2$ has the same structure as $T_1$ except that

- all sequences $S_i$ and $S_{i,j}$ are sorted.
- each circuit in Figure 13 (i) is changed to (ii).

In order to transform $T_1$ into $T_2$, we need to sort the sequences $S_i$ and $S_{i,j}$ and convert each one-way circuit to the structure shown in Figure 13 (ii). If the set $S$ has an exact cover $C_{i_1},\dots,C_{i_q}$, we can do the transformation efficiently as follows. For each $C_j=\{s_{j_1},s_{j_2},s_{j_2}\}$, $(j=i_1,\dots,i_q)$, in the cover, we send the three sequences $S_{j_1},S_{j_2},S_{j_3}$ on the left of $T_1$ to their counterparts $S_{j,j_1},S_{j,j_2},S_{j,j_3}$ on the right of $T_1$, combine the sorting of each pair as explained in Figure 12, then move the sorted $S_{j_1},S_{j_2},S_{j_3}$ sequences back. During this process we also get each one-way circuit involved into the correct shape. We then sort the other un-sorted sequences $S_{i,j}$ on the right of $T_1$ and get their leading one-way circuits into the correct shape.

The total cost $N$ for this process is calculated as follows. Recall that we send precisely $q$ groups of sequences from the left to the right of $T_1$.

1. The overhead for these $q$ groups to cross the tree connection network: $q(\log m+\log n)+O(1)$ nni moves. This is done by grouping three sequences in each group first, then moving them as one unit.
2. The cost of crossing the $q$ toll sequences of length $m^2$ before the first batch of one-way circuits: $qm^2$ nni moves.

3. Converting each one-way circuit to the structure in Figure 13 (ii) costs $7r-7$ nni moves.
4. Moving a group of sequences across a one-way circuit and back costs $O(1)$ extra nni moves, for each of the $q$ groups.
5. The cost of sorting $3q$ pairs of sequences (by combining).
6. the cost of sorting the remaining $3n - 3q$ sequences individually.

The following lemma completes the reduction and thus the proof of Theorem 1.

LEMMA 10. *If the set $S$ has no exact cover, then $D_{nni}(T_1, T_2) \geq N + m^2/2$.*

PROOF. Suppose that $S$ has no exact cover. Then to transform $T_1$ into $T_2$, we either have to send more than $q$ groups (some groups with less than 3 sequences) to the right crossing the one-way circuits, or some sequence $S_i$ is sorted separately from $S_{j,i}$'s or some sequence $S_i$ is sorted together with a "wrong" sequence $S_{j,h}$, where $h \neq i$. In the first case, the cost will be increased by $m^2$ nni moves, which is the cost of moving an extra group past a delimiter sequence of length $m^2$. In the last case, at least one segment of $m^3$ $x$'s is sorted together with a segment of $y$'s. By Lemma 9 and the choice $\epsilon = \frac{1}{8}$, this is not much better than sorting the two segments separately and costs at least $(c - \frac{1}{8})m^3 k \log k - m^3 k$ more nni moves than sorting one such segment, which is larger than $m^2$ for sufficiently large $k$ and $m$, since $c > \frac{1}{4}$. The second case introduces an extra cost of $q_i - 12km^3 \log m - m^2 = \Omega((m^3 \log m) k \log k)$ by equation (5) and Lemma 9, which is again larger than $m^2$ for sufficiently large $k$ and $m$.

Notice that in the above definition of $N$, the bounds in items 2,3,6 are all optimal. The bounds in items 1 and 5 are the worst case overheads and may not be optimal. But these two items only account for $O(m(\log m + \log n))$ nni moves, which is not sufficient to compensate for the extra cost $m^2$ given above. □

DEPARTMENT OF COMPUTER SCIENCE, RUTGERS UNIVERSITY, CAMDEN, NJ 08102, USA
*E-mail address*: bhaskar@crab.rutgers.edu

DEPARTMENT OF COMPUTER SCIENCE, SUNY-BUFFALO, BUFFALO, NY 14260, USA
*E-mail address*: xinhe@cs.buffalo.edu

DEPARTMENT OF COMPUTER SCIENCE, UNIVERSITY OF CALIFORNIA AR RIVERSIDE, RIVERSIDE, CA 92521, USA
*E-mail address*: jiang@cs.ucr.edu

DEPARTMENT OF COMPUTER SCIENCE, UNIVERSITY OF WATERLOO, WATERLOO, ONTARIO N2L 3G1, CANADA
*E-mail address*: mli@math.uwaterloo.ca

CWI, P.O. BOX 94079, 1090 GB AMSTERDAM, NETHERLANDS
*E-mail address*: tromp@cwi.nl

BIOINFORMATICS CENTER, KENT RIDGE DIGITAL LABS, HENG MUI KENG TERRACE, SINGAPORE 119597
*E-mail address*: lxzhang@bic.nus.edu.sg

DIMACS Series in Discrete Mathematics
and Theoretical Computer Science
Volume **55**, 2000

# Two-Stage Group Testing for Complexes in the Presence of Errors

Anthony J. Macula, David C. Torney, and Pavel A.Vilenkin

ABSTRACT: The screening of biological data sets is an essential aspect of modern biotechnology. Whenever the objective is to find positive objects (e.g., molecules), a test indicating whether at least one positive is in a specific part of the data set can greatly facilitate their isolation. Such tests are called *binary group tests* and the general mathematical method behind the identification of the positives using such tests is known as *classical group testing*. In several biotechnical settings, the use of classical group testing to isolate objects which are *individually* positive has become standard experimental procedure. However, very little work has been done in applying group testing techinques to the identication of objects which are *collectively* positve. In this paper, we discuss a probabilistic group testing method which leads to the identification of *positive combinations of objects*.

Let [t] represent a finite population with t elements. Suppose we have an unknown collection of d+1 k-subsets C of [t]. We refer to C as the set of *positive k-complexes* and we simply call a subset in C a *k-complex* In the *generalized group testing problem*, C must be identified by performing 0 , 1 tests on subsets or *pools* of [t]. A pool is said to be positive if it completely contains a k-complex; otherwise the pool is said to be negative. In classical group testing, each member of C is a singleton or a 1-complex. In this paper, we exhibit a pseudo-random two-stage algorithm that identifies a large proportion of the positive complexes in the presence of errors.

## 1. Group testing and biological screening

The screening of biological data sets is an essential, but often daunting aspect of modern biotechnology. So daunting that, without efficient screening methods, the desired experimental analysis would be impractical. Whenever the objective is to find "positive objects", a test indicating whether at least one positive is in a specific part of the data set can greatly facilitate the isolation of the positives. Such tests are called *binary group tests* and the general mathematical method behind the identification of the positives using such tests is known as *classical group testing*. In several biotechnical settings, the application of classical group testing to the screening of large biological data sets (e.g., cDNA libraries) that contain a few number of data points which have an attribute characterizing them a being positive (e.g., clones positive for some PCR reaction) has become standard experimental procedure. See [2] , [3], [5], [6], [7], [13],

*2000 Mathematics Subject Classification.* Primary 62K99 92-08; Secondary 05B20, 05D05, 62P10.
*Keywords.* group testing, two-stage algorithm, postive combinations, complexes, sets pooling.
The first author was partially supported by NSF grant DMS-997352.

and [14]. Given a finite ground set, data set, or *population* containing elements which can be uniquely characterized as positive or negative, in classical group testing, the positives are identified by performing 0 , 1 tests on subsets the population. In this paper, we consider a biologically motivated generalization of the classical group testing problem because:

*The molecules of biological cells are being identified, but it remains a challenge to understand how they cooperate to produce various attributes of biological cells [14].*

This quote leads one to consider the problem of *group testing for complexes*. A *pool* is a subset of a biological data set. A *complex pool assay* is a 0 , 1 test to determine if a pool contains one of possibly several different *combinations of data points* that produce or manifest some measurable function or property. A complex pooling design is a collection of pools. A practical goal is to determine a portion of the positive combinations or positive complexes from the pool assays. In this paper, we consider a constrained version of this general problem. For simplicity, we focus on combinations of a fixed size k. However, using ideas like those contained here and in [9], our methods can be generalized to arbitrary combinations.

Group testing for complexes has many potential applications. Whenever certain positive combinations are the object of a search, our method could be applied whenever there is a way to detect the presence of a positive combination in a pool.

Consider the following example which could be a model for a genetic experiment. Suppose we have a population of binary sequences of length 20. Some are normal, some are "diseased". Suppose, unbeknownst to the experimenter, that a sequence is diseased if and only if it has 1s in the first, seventh, and eleventh entries {1, 7, 11} or, has 1s in its seventh, tenth or nineteenth entries {7, 10, 19}. Suppose the experimenter only knows a diseased sequence phenotypically and is trying to find out which combinations of 1s actually give rise to the disease. Take one diseased sequence and parse it into several subsequences in a prescribed way. Then compare the subsequences of the diseased sequence to a database of identically parsed normal sequences. If one particular subsequence of the diseased sequence does not match any of the corresponding normal subsequences, then one can assume that the diseased subsequence contains a combination of 1s that gives rise to the disease. In particular, any disease subsequence that includes the first, seventh, and eleven entries, or the seventh, tenth, and nineteenth entries can not possibly match any corresponding normal subsequence.

This experiment fits the group testing for complexes paradigm in the following way: The population is the set entries of the of disease sequence. In this case, the set {1, ... , 20}. The pools are the subsequences. The positive complexes are the subsets {1, 7, 11} and {7, 10, 19}. A subsequence is positive if it doesn't match any normal subsequence.

## 2. Group testing for complexes

Throughout the remainder of this paper, all simple lower case variables are non-negative integers. Given set S , |S| denotes its cardinality. [t] denotes the positive integers $\{1, 2, \ldots, t\}$. A subset of $[t]$ with cardinality $k$ is called a *k-set.* $\binom{[t]}{k}$ denotes the k-sets of $[t]$.

Let $[t]$ represent a finite population with $t$ elements. Suppose we have an unknown collection of $d + 1$ k-subsets $C$ of $[t]$. We refer to $C$ as the set of *positive k-complexes* and we simply call a subset in $C$ a *k-complex.* In the *generalized group testing problem,* $C$ must be identified by performing $0$, $1$ tests on subsets or pools of $[t]$. A pool is said to be positive (1) if it *completely contains* a k-complex; otherwise the pool is said to be negative (0). In classical group testing, each member of $C$ is a singleton or a 1-complex. In [14], a random pooling design was used to address this problem. In this paper, we discuss a pseudo-random approach to this problem. Our approach enables us to get a lower bound on the performance of our designs in the presence of errors.

## 3. Random and pseudo-random matrices

DEFINITION 1. For $0 < p < 1$, we let $M(n,p,t)$ be a random $n \times t$ $0$, $1$ matrix, each entry of which, is $1$ with probability $p$. For $1 \leq x \leq n$, let $\mathbf{r}_x$ denote the xth row of $M(n,p,t)$.

Fix an $M(n,p,t)$. We can identify an element $u$ of $[t]$ with the uth column of this fixed $M(n,p,t)$. We can in turn identify the uth column of $M(n,p,t)$ with the subset $\left\{ x \in [n] : r_x \text{ has a } 1 \text{ in the uth column} \right\}$. Thus, using a $M(n,p,t)$, we can identify the elements of $[t]$ with subsets of $[n]$. We call this the *$M(n,p,t)$ representation of $u \in [t]$.* See FIGURE 1.

DEFINITION 2. Given an $M(n,p,t)$, we define the $\binom{n}{2} \times t$ matrix $M^2(n,p,t)$ by setting row $\mathbf{r}_{\{x,y\}}$, where $\{x,y\} \subset [n]$, to be the intersection of $\mathbf{r}_x$ and $\mathbf{r}_y$ in $M(n,p,t)$. For simplicity, we just write $\mathbf{r}_{xy}$ or $\mathbf{r}_{yx}$ to indicate $\mathbf{r}_{\{x,y\}}$. We use the same labels for the columns of $M^2(n,p,t)$ as we did for $M(n,p,t)$. *Note, we do not think of the columns of $M^2(n,p,t)$ as subsets of $\left[\binom{n}{2}\right]$.* A column of $M^2(n,p,t)$ is represented as a subset of $[n]$ using the $M(n,p,t)$ representation. See FIGURE 2.

| $[n] \equiv$ | $\{x,y,w\}$ | $\{z,w\}$ | $\{y,z,w\}$ | $\{x,y,z\}$ | $\{x,y\}$ | $\{x,y,z,w\}$ | $\{z,w\}$ $\equiv [t]$ |
|---|---|---|---|---|---|---|---|
| x | 1 | 0 | 0 | 1 | 1 | 1 | 0 |
| y | 1 | 0 | 1 | 1 | 1 | 1 | 0 |
| z | 0 | 1 | 1 | 1 | 0 | 1 | 1 |
| w | 1 | 1 | 1 | 0 | 0 | 1 | 1 |

FIGURE 1
M(4, 0.75, 7)

| | $\{x,y,w\}$ | $\{z,w\}$ | $\{y,z,w\}$ | $\{x,y,z\}$ | $\{x,y\}$ | $\{x,y,z,w\}$ | $\{z,w\}$ $\equiv [t]$ |
|---|---|---|---|---|---|---|---|
| xy | 1 | 0 | 0 | 1 | 1 | 1 | 0 |
| xz | 0 | 0 | 0 | 1 | 0 | 1 | 0 |
| xw | 1 | 0 | 0 | 0 | 0 | 1 | 0 |
| yz | 0 | 0 | 1 | 1 | 0 | 1 | 0 |
| yw | 1 | 0 | 1 | 0 | 0 | 1 | 0 |
| zw | 0 | 1 | 1 | 0 | 0 | 1 | 1 |

FIGURE 2
$M^2(4, 0.75, 7)$

Henceforth, unless otherwise specified, assume the uniform distribution on [t] and $\binom{[t]}{k}$. Using the M(n,p,t) representation, we think of each $u \in [t]$ as a subset of [n]. Let $S_i \subset [t]$ be a k-set. Because [t] is identified with the columns of M(n,p,t) (which, in turn, are identified with the columns of $M^2(n,p,t)$ ), we can think of each $S_i$ as a k-set of columns in M(n,p,t) (or $M^2(n,p,t)$). Thus a k-complex in [t] can be identified with a k-complex in the power set of [n]. Then for $S_i \subset [t]$ , we write $\bigcap S_i$ to denote the intersection in [n] of the M(n,p,t) representation of columns in $S_i$. See FIGURE 3a.

$$\begin{Bmatrix}x\\y\\w\end{Bmatrix}\begin{Bmatrix}z\\w\end{Bmatrix}\begin{Bmatrix}y\\z\\w\end{Bmatrix}\begin{Bmatrix}x\\y\\z\end{Bmatrix}\begin{Bmatrix}x\\y\end{Bmatrix}\begin{Bmatrix}x\\y\\z\\w\end{Bmatrix}\begin{Bmatrix}z\\w\end{Bmatrix} \equiv \{u_0 , \dots , u_6\}$$

$[n] \equiv$

| | | | | | | | |
|---|---|---|---|---|---|---|---|
| x | 1 | 0 | 0 | 1 | 1 | 1 | 0 |
| y | 1 | 0 | 1 | 1 | 1 | 1 | 0 |
| z | 0 | 1 | 1 | 1 | 0 | 1 | 1 |
| w | 1 | 1 | 1 | 0 | 0 | 1 | 1 |

$$S_0 = \{u_0, u_2, u_5\},\ S_1 = \{u_1, u_3, u_6\} \Rightarrow S_0 = \left\{\begin{Bmatrix}x\\y\\w\end{Bmatrix},\begin{Bmatrix}y\\z\\w\end{Bmatrix},\begin{Bmatrix}x\\y\\z\\w\end{Bmatrix}\right\},\ S_1 = \left\{\begin{Bmatrix}z\\w\end{Bmatrix},\begin{Bmatrix}x\\y\\z\end{Bmatrix},\begin{Bmatrix}z\\w\end{Bmatrix}\right\}$$

$$\cap S_0 = \{y, w\} \qquad\qquad \cap S_1 = \{z\}$$

FIGURE 3a

Now, identifying $[t]$ with the columns of $M^2(n,p,t)$ , a row $\mathbf{r}_{xy}$ of $M^2(n,p,t)$ determines a subset of $[t]$. That is, a given element $u \in [t]$ is in the subset determined by $\mathbf{r}_{xy}$ if and only if there is a 1 in the entry where $\mathbf{r}_{xy}$ , and the column which corresponds to $u$ , intersect.

$$\begin{Bmatrix}x\\y\\w\end{Bmatrix}\begin{Bmatrix}z\\w\end{Bmatrix}\begin{Bmatrix}y\\z\\w\end{Bmatrix}\begin{Bmatrix}x\\y\\z\end{Bmatrix}\begin{Bmatrix}x\\y\end{Bmatrix}\begin{Bmatrix}x\\y\\z\\w\end{Bmatrix}\begin{Bmatrix}z\\w\end{Bmatrix} \equiv \{u_0 , \dots , u_6\}$$

| | | | | | | | |
|---|---|---|---|---|---|---|---|
| xy | 1 | 0 | 0 | 1 | 1 | 1 | 0 |
| xz | 0 | 0 | 0 | 1 | 0 | 1 | 0 |
| xw | 1 | 0 | 0 | 0 | 0 | 1 | 0 |
| yz | 0 | 0 | 1 | 1 | 0 | 1 | 0 |
| yw | 1 | 0 | 1 | 0 | 0 | 1 | 0 |
| zw | 0 | 1 | 1 | 0 | 0 | 1 | 1 |

$$\mathbf{r}_{xy} = \{u_0 , u_3 , u_4 , u_5\} = \left\{\begin{Bmatrix}x\\y\\w\end{Bmatrix},\begin{Bmatrix}x\\y\\z\end{Bmatrix},\begin{Bmatrix}x\\y\end{Bmatrix},\begin{Bmatrix}x\\y\\z\\w\end{Bmatrix}\right\}$$

FIGURE 3b

In short, the subset $\mathbf{r}_{xy}$ is the collection of all $u \in [t]$ with $\{x,y\} \subset u$ under the M(n,p,t) representation. See FIGURE 3b. The subsets $\mathbf{r}_{xy}$ will be the pools in our design. See Section 4.

DEFINITION 3. Let $0 < k < t$ and let $u_0, u_1, \ldots, u_k \in [t]$ be $k+1$ randomly selected elements without replacement and let $S_0, S_1, \ldots, S_d$, with $S_i \subset [t]$ for $0 \le i \le d$, be $d+1$ randomly selected k-complexes with replacement. Let $0 < p < 1$, $1 \le m \le n$, and $0 < d < \binom{t}{k}$. We define $\Phi(m,n,d,k,t,p)$ to be the probability that *there is* a row $\mathbf{r}_x$ in M(n,p,t) with $1 \le x \le m$ that has a 1 in *each* column contained in $S_0$ and, for each $1 \le i \le d$, a 0 in *some* column of $S_i$. We define $\gamma(n,k,t,p)$ to be the probability that *there isn't* a row $\mathbf{r}_x$ in M(n,p,t) with $1 \le x \le n$ that has a 0 in the $u_0$ column and 1s in each of the other $u_i$ columns for $1 \le i \le k$.

If we think of the columns of M(n,p,t) as subsets of [n], then $\gamma(n,k,t,p)$ is the probability that *there isn't* an $x \in [n]$ with $x \notin u_0$ and $x \in u_1$ for $1 \le i \le k$ and $\Phi(m,n,d,k,t,p)$ is the probability that, for $1 \le i \le d$, *there is* an $x \in [m]$ with $x \in \cap S_0$ and $x \notin \cap S_i$.

**PROPOSITION 1.** *Let $q = 1 - p$. Then $\Phi(m,n,d,k,t,p) \ge$*

$$m \sum_{w=0}^{t} \binom{t}{w} p^w q^{t-w} \binom{w}{k}\binom{t}{k}^{-1}\left(1 - \binom{w}{k}\binom{t}{k}^{-1}\right)^d -$$

$$\binom{m}{2} \sum_{\substack{h,i,j,\\ i+j+h \le t}} \binom{t}{h}\binom{t-h}{i}\binom{t-h-i}{j} p^{2h}(pq)^{i+j} q^{2(t-i-j-h)} \frac{\binom{h}{k}\left(\binom{t}{k} - \binom{h+i}{k} - \binom{h+j}{k} + \binom{h}{k}\right)^d}{\binom{t}{k}^{(d+1)}}.$$

PROOF: Let $S_0, S_1, \ldots, S_d$, with $S_i \subset [t]$ for $0 \le i \le d$, be $d+1$ randomly selected k-complexes with replacement. Let $\mathbf{r}_x$ be a row of M(n,p,t). The weight of that row is $w$ with probability $\binom{t}{w} p^w q^{t-w}$. Let $F_x$ be the event that $\mathbf{r}_x$ contains $S_0$ and doesn't contain $S_i$ for $1 \le i \le d$. Then $\mathrm{Prob}(F_x) =$

$\sum_{w=0}^{t} \binom{t}{w} p^w q^{t-w} \binom{w}{k}\binom{t}{k}^{-1}\left(1 - \binom{w}{k}\binom{t}{k}^{-1}\right)^d$. We have that $\Phi(m,n,d,k,t,p) =$

$\mathrm{Prob}\left(\bigcup_{x=1}^{m} F_x\right) \ge m \cdot \mathrm{Prob}(F_1) - \binom{m}{2} \mathrm{Prob}(F_1 \cap F_2)$. Since $\mathrm{Prob}(F_1 \cap F_2) =$

$$\sum_{\substack{h,i,j,\\ i+j+h\le t}} \binom{t}{h}\binom{t-h}{i}\binom{t-h-i}{j} p^{2h}(pq)^{i+j} q^{2(t-i-j-h)} \frac{\left(\binom{h}{k}\left(\binom{t}{k}-\binom{h+i}{k}-\binom{h+j}{k}+\binom{h}{k}\right)\right)^{d}}{\binom{t}{k}^{(d+1)}},$$

the result follows. Q. E. D.

DEFINITION 4. Let $k(d+1) \le t$ and let $S_0, S_1, \ldots, S_d$, with $S_i \subset [t]$ for $0 \le i \le d$, be a fixed collection of $d+1$ *disjoint* k-complexes. We define $\Phi_0(m, n, d, k, t, p)$ to be the probability that there is a row $\mathbf{r}_x$ in M(n,p,t) with $1 \le x \le m$ that has a 1 in each column contained in $S_0$ and, for each $1 \le i \le d$, a 0 in some column of $S_i$.

**PROPOSITION 2.** $\Phi_0(m, n, d, k, t, p) = 1-(1-p^k(1-p^k)^d)^m$.

PROOF: Let $k(d+1) \le t$ and let $\mathbf{r}_x$ be a row of M(n,p,t). Let $E_x$ be the event that $\mathbf{r}_x$ contains $S_0$ and doesn't contain $S_i$ for $1 \le i \le d$. Then $\mathrm{Prob}(E_x) = p^k(1-p^k)^d$. Since the events $\{E_x\}_{x\in[m]}$ are independent, we have that $\mathrm{Prob}\left(\bigcup_{x=1}^{m} E_x\right) =$

$$1-\mathrm{Prob}\left(\bigcap_{x=1}^{m} \neg E_x\right) = 1-(1-p^k(1-p^k)^d)^m.$$ Q. E. D.

**PROPOSITION 3.** *Let $q = 1 - p$. Then* $\gamma(n,k,t,p) = \left(1-qp^k\right)^n$.

PROOF: Suppose we select $k + 1$ columns $u_0, u_1, \ldots, u_k$ of M(n,p,t) without replacement. The probability that a given row of M(n,p,t) has a 0 in the $u_0$ column and 1s in each of the other $u_i$ columns for $1 \le i \le k$ is $qp^k$. Since the rows are independently generated, the probability that there isn't a row of M(n,p,t) with a 0 in the $u_0$ column and 1s in each of the other $u_i$ columns for $0 \le i \le k$ is $\gamma(n,k,t,p) = \left(1-qp^k\right)^n$. Q.E.D.

DEFINITION 5. For a given M(n,p,t), let $S_1$ and $S_2$ be two k-complexes randomly selected from [t] with replacement. Let $\lambda(n,k,p,t)$ be the probability that $\bigcap S_1 = \bigcap S_2$. If $S_1$ and $S_2$ are two disjoint k-complexes in [t], let $\lambda_0(n,k,p,t)$ be the probability that $\bigcap S_1 = \bigcap S_2$.

**PROPOSITION 4.**

*a.* $$\lambda(n,k,p,t) = \sum_{r=0}^{k} \binom{k}{r}\binom{t-k}{k-r}\binom{t}{k}^{-1} \left((1-p^r) + p^r\left(p^{(2k-2r)} + (1-p^{k-r})^2\right)\right)^n.$$

*b.* $$\lambda_0(n,k,p,t) = \left(p^{2k} + (1-p^k)^2\right)^n.$$

PROOF: We give the proof of a. Then b essentially follows from a. Suppose $C \subset [t]$ is the set of columns that $S_1$ and $S_2$ have in common. Suppose $|C| = r$. The probability that $|\bigcap C| = s$ is $\binom{n}{s} p^{rs}(1-p^r)^{n-s}$. Let $I = \bigcap C$. Now $\bigcap S_1 = \bigcap S_2$ if and only if $I \cap \bigcap(S_1 \setminus C) = I \cap \bigcap(S_2 \setminus C)$. Suppose $\bigcap S_1 = \bigcap S_2$. Then for each $x \in I$, either $x \in \bigcap S_i$ or $x \notin \bigcap S_i$ for $i = 1, 2$. Let this event be $H_x$. Since $|S_i \setminus C| = k - r$ for $i = 1, 2$, then $\text{Prob}(H_x) = p^{2(k-r)} + (1 - p^{k-r})^2$. Since the rows are independently generated and $\text{Prob}(|C| = r) = \binom{t-k}{k-r}\binom{k}{r}\binom{t}{k}^{-1}$, we have that $\lambda(n,k,p,t)$

$$= \sum_{r=0}^{k}\sum_{s=0}^{n} \binom{k}{r}\binom{t-k}{k-r}\binom{t}{k}^{-1}\binom{n}{s}(1-p^r)^{n-s}p^{rs}\left(p^{(2k-2r)} + (1-p^{k-r})^2\right)^s$$

$$= \sum_{r=0}^{k} \binom{k}{r}\binom{t-k}{k-r}\binom{t}{k}^{-1}\left((1-p^r) + p^r\left(p^{(2k-2r)} + (1-p^{k-r})^2\right)\right)^n.$$

## 4. Pooling

Select an $M(n,p,t)$ and use it to construct $M^2(n,p,t)$. We use the matrix $M^2(n,p,t)$ to construct our pools. We identify a population of cardinality $[t]$ with the columns of $M^2(n,p,t)$. Then a row $\mathbf{r}_{xy}$ of $M^2(n,p,t)$ determines a pool of the population in the obvious way. That is, a given element $u \in [t]$ is in the pool determined by $\mathbf{r}_{xy}$ if and only if there is a 1 in the entry where $\mathbf{r}_{xy}$, and the column which corresponds to $u$, intersect. In short, the pool $\mathbf{r}_{xy}$ is the collection of all $u \in [t]$ with $\{x,y\} \subset u$ under the $M(n,p,t)$ representation. Suppose our positive k-complexes, $S_0, S_1, \ldots, S_d$ in $[t]$, are randomly selected *without* replacement. However, for all practical purposes, we can assume that they are selected *with* replacement. Then each $S_i$ for $0 \le i \le d$ is a k-complex $\{u_{i1}, \ldots, u_{ik}\}$ where $u_{ij} \subset [n]$. By testing pool $\mathbf{r}_{xy}$, we define an *output vector* $\mathbf{o}$ by setting entry $\mathbf{o}_{xy}$ equal to 1 if the test result of pool $\mathbf{r}_{xy}$ is positive and 0 if not. In other words, pool $\mathbf{r}_{xy}$ is positive if and only if $\{x,y\} \subset \bigcap_{j=1}^{k} u_{ij}$ for some $0 \le i \le d$. Note $\mathbf{o}_{xy}$ and $\mathbf{o}_{yx}$ mean the same thing.

## 5. Complexes decoding algorithm

Suppose we have d + 1 positive k-complexes $S_0, S_1, \dots, S_d$ in a populations [t] and we have an output vector **o** generated by an application of the pooling strategy outlined above.

STAGE 1. 1. For each x in [n] , take the set $\{x\} \cup \{y: \mathbf{o}_{xy} = 1\} \subset [n]$. Call this set $I_x$ the *intersection pivot at x.*

STAGE 1. 2. Let $\mathbf{b}_x$ be the binary representation of $I_x \subset [n]$. Call $\mathbf{b}_x$ the *pivot vector* at x. Consider the set of all columns of M(n,p,t) which are *equal to or above* $\mathbf{b}_x$ in the coordinate-wise order. Call this collection of the columns $B_x$ the *pivot union at x.*

STAGE 1. 3. Consider the k-complexes of columns of $B_x$ which have $\mathbf{b}_x$ as their intersection. Call this collection of k-complexes $C_x$ the *candidate k-complexes at x.*

STAGE 2. For each $x \in [n]$, test each candidate k-complex of $C_x$ individually.

EXAMPLE 1. In FIGURES 4a and 4b below, we use a small example to demonstrate the pooling and decoding scheme for a M(5,0.75,8). Here the population is $\{u_0, u_1, \dots, u_7\}$. The positive 2-complexes are $S_0 = \{u_0, u_1\}$, $S_1 = \{u_0, u_6\}$, $S_2 = \{u_3, u_6\}$. Notice that $B_2 = S_1$ and $S_2 \subset B_3$.

$Population = \{u_0, u_1, u_2, u_3, u_4, u_5, u_6, u_7\}$

$S_0 = \{u_0, u_1\}, S_1 = \{u_0, u_6\}, S_2 = \{u_3, u_6\}$

| $M(5,0.75,8)$ | $u_0$ $\{1,3,4,5\}$ | $u_1$ $\{1,2,3,5\}$ | $u_2$ $\{2,3,4,5\}$ | $u_3$ $\{1,3,4\}$ | $u_4$ $\{2,3,4,5\}$ | $u_5$ $\{1,2,3,5\}$ | $u_6$ $\{1,2,4,5\}$ | $u_7$ $\{1,3,4\}$ |
|---|---|---|---|---|---|---|---|---|
| 1 | 1 | 1 | 0 | 1 | 0 | 1 | 1 | 1 |
| 2 | 0 | 1 | 1 | 1 | 1 | 1 | 1 | 0 |
| 3 | 1 | 1 | 1 | 0 | 1 | 1 | 0 | 1 |
| 4 | 1 | 0 | 1 | 1 | 1 | 0 | 1 | 1 |
| 5 | 1 | 1 | 1 | 0 | 1 | 1 | 1 | 0 |

$\bigcap S_0 = \{1,3,5\}$ $\bigcap S_1 = \{1,4,5\}$ $\bigcap S_2 = \{1,2,4\}$

Figure 4a

| $M^2(5,0.75,8)$ | $u_0$ | $u_1$ | $u_2$ | $u_3$ | $u_4$ | $u_5$ | $u_6$ | $u_7$ | pool outcomes | outcome vector |
|---|---|---|---|---|---|---|---|---|---|---|
| 12 | 0 | 1 | 0 | 1 | 0 | 1 | 1 | 0 | + | $o_{12}=1$ |
| 13 | 1 | 1 | 0 | 0 | 0 | 1 | 0 | 1 | + | $o_{13}=1$ |
| 14 | 1 | 0 | 0 | 1 | 0 | 0 | 1 | 1 | + | $o_{14}=1$ |
| 15 | 1 | 1 | 0 | 0 | 0 | 1 | 1 | 0 | + | $o_{15}=1$ |
| 23 | 0 | 1 | 1 | 0 | 1 | 1 | 0 | 0 | − | $o_{23}=0$ |
| 24 | 0 | 0 | 1 | 1 | 1 | 0 | 1 | 0 | + | $o_{24}=1$ |
| 25 | 0 | 1 | 1 | 0 | 1 | 1 | 1 | 0 | − | $o_{25}=0$ |
| 34 | 1 | 0 | 1 | 0 | 1 | 0 | 0 | 1 | − | $o_{34}=0$ |
| 35 | 1 | 1 | 1 | 0 | 1 | 1 | 0 | 0 | + | $o_{35}=1$ |
| 45 | 1 | 0 | 1 | 0 | 1 | 0 | 1 | 0 | + | $o_{45}=1$ |

$I_1=\{1,2,3,4,5\}$ $I_2=\{1,2,4\}$ $I_3=\{1,3,5\}$ $I_4=\{1,2,4,5\}$ $I_5=\{1,3,4,5\}$

$$b_1=\begin{pmatrix}1\\1\\1\\1\\1\end{pmatrix}\quad b_2=\begin{pmatrix}1\\1\\0\\1\\0\end{pmatrix}\quad b_3=\begin{pmatrix}1\\0\\1\\0\\1\end{pmatrix}\quad b_4=\begin{pmatrix}1\\1\\0\\1\\1\end{pmatrix}\quad b_5=\begin{pmatrix}1\\0\\1\\1\\1\end{pmatrix}$$

$B_1=\varnothing$ $B_2=\{u_3,u_6\}$ $B_3=\{u_0,u_1,u_5\}$ $B_4=\{u_6\}$ $B_5=\{u_0\}$

FIGURE 4b

## 6. Analysis of the method

Because $\mathbf{o}_{xy}=1$ if and only if $\{x,y\}\subset\bigcap_{j=1}^{k}u_{ij}$, it is straightforward to verify that $I_x=\cap S_i=\bigcap_{j=1}^{k}u_{ij}$ if $|\cap S_i|\geq 2$ and $x\in\cap S_i$ and $x\notin\cap S_{i'}$ for each $i\neq i'$. Since in most of our applications the, $\text{Prob}(|\cap S_i|\geq 2)\approx 1$, we can conclude that $\Phi(n,n,d,k,t,p)$ is the probability that, for a given $S_i$, there is $x\in[n]$ with $I_x=\cap S_i=\bigcap_{j=1}^{k}u_{ij}$. Thus of the $d+1$ positive k-complexes, an expected number of $(d+1)\cdot\Phi(n,n,d,k,t,p)$ of them have their intersection equal to an intersection pivot for some x.

Consider $B_x$. Suppose $I_x = \cap S_i = \bigcap_{j=1}^{k} u_{ij}$ for some $S_i$. Then the expected number of columns of $M(n,p,t)$ which are equal to or above $\mathbf{b}_x$, i.e., the expected value of $|B_x|$, is $k + (t-k)\cdot\gamma(n,k,t,p)$. Obviously, if $I_x = \bigcap_{j=1}^{k} u_{ij}$, then $S_i \subset B_x$ because $\mathbf{b}_x$ is the intersection of the $k$ columns $u_{i1}, \ldots, u_{ik}$. Suppose a column $u_0 \notin \{u_{i1}, \ldots, u_{ik}\}$. The result follows from the observation that $u_0$ is equal to or above $\mathbf{b}_x$ if an only if there isn't a row of $M(n,p,t)$ that has a 0 in the $u_0$ column and 1s in each of the columns $u_{i1}, \ldots, u_{ik}$. Since $\gamma(n,k,t,p)$ is the probability that $u_0$ has this property, the expected number of such $u_0$ that are equal to, or above, $\mathbf{b}_x$ is $(t-k)\cdot\gamma(n,k,t,p)$.

Consider $C_x$ and suppose $I_x = \cap S_i = \bigcap_{j=1}^{k} u_{ij}$ for some $S_i$. Since $\lambda(n,k,p,t)$ is the probability that two k-complexes have the same intersection, then the expected value of $|C_x|$ is $\binom{t}{k}\lambda(n,k,p,t)$. If we use an $M^2(n,p,t)$ pooling strategy, then $\binom{n}{2}$ first stage tests identify an expected number of $(d+1)\cdot\Phi(n,n,d,k,t,p)$ positive k-complexes with intersection equal to one of the $n$ intersection pivots. For each intersection pivot, $\binom{t}{k}\lambda(n,k,p,t)$ confirmatory tests are required. Thus we have the following result:

**PROPOSITION 5.** *For a population of size $t$ with $d+1$ positive k-complexes, in the absence of errors, using an $M^2(n,p,t)$ pooling design and decoding strategy, an expected number of $\binom{n}{2}+n\binom{t}{k}\lambda(n,k,p,t)$ tests identify an expected number of $(d+1)\cdot\Phi(n,n,d,k,t,p)$ positive k-complexes. If the $d+1$ positive k-complexes are disjoint, then an expected number of $\binom{n}{2}+n\binom{t}{k}\lambda_0(n,k,p,t)$ tests identify an expected number of $(d+1)\cdot\Phi_0(n,n,d,k,t,p)$ positive k-complexes.*

EXAMPLE 2. Suppose $t = 100$ and there are six positive and disjoint 4-complexes, i.e., $d = 5$, $k = 4$. If we use an $M^2(32,0.74,100)$ pooling design and decoding strategy, then, using PROPOSITION 2, an expected number of 5.05 positive 4-complexes are identified using an expected number of 511 tests.

## 7. Testing with errors

Here we take the most straightforward analysis of performance in the presence of errors. If pool $\mathbf{r}_{xy}$ contains a positive k-complex, but is mislabeled by the test result as a negative pool or vice-versa, then we say that $\mathbf{o}_{xy}$ is an *outcome error.*

How do outcome errors affect the efficacy of our method? Because our method relies on an assumed error-free confirmatory stage, errors will only effect the performance of our method and will not lead to the identification of false positives. Suppose we use an $M^2(n,p,t)$ pooling design and decoding strategy. Suppose our positive k-complexes are $S_0$ , $S_1$ ,... , $S_d$ in $[t]$ and that $e$ errors occurred in the testing. Consider $S_0 = \{u_{01}, \ldots, u_{0k}\}$. What is the probability that there is still an $x \in [n]$ with $I_x = \cap S_0 = \bigcap_{j=1}^{k} u_{0j}$ ? Let $E = \bigcup\left\{\{x,y\} : \mathbf{o}_{xy} \text{ is and outcome error}\right\}$. If there is an $x_0 \in [n]-E$ with $x_0 \in \cap S_0$ and $x_0 \notin \cap S_i$ for $i \neq 0$ and $|\cap S_0| \geq 2$ , then $I_{x_0} = \bigcap_{j=1}^{k} S_{0j}$ regardless of the outcome errors because no outcome of the form $\mathbf{o}_{x_0 y}$ is an error and $\mathbf{o}_{x_0 y} = 1$ if and only if $\{x_0, y\} \subset \bigcap S_0$. Thus if $|E| = e$ , then $\Phi(n-e,n,d,k,t,p)$ is approximately the probability that there is an $x_0 \in [n]-E$ with $x_0 \in \cap S_0$ and $x_0 \notin \cap S_i$ for $i \neq 0$ and $|\cap S_0| \geq 2$ .

If $Y(e)$ is the random variable that gives the cardinality of the union of $e$ 2-sets of $[n]$ (with repetition allowed), then the expected number k-complexes with intersection equal to some $I_x$ is approximately equal to $(d+1)\cdot\Phi(n-Y(e),n,d,k,t,p)$.

Since $\operatorname{Prob}(Y(e)=y) = \binom{n}{2}^{-e}\binom{n}{y}\sum_{i=0}^{\min(n,2e)}(-1)^i\binom{y}{i}\binom{y-i}{2}^e$ , we have:

**THEOREM 1.** *If there are d + 1 positive k-complexes $S_0$ , $S_1$ ,... , $S_d$ in a population [t] and e errors occurred during the testing of the first stage pools generated by $M^2(n,p,t)$, then the expected number of k-complexes with intersection equal to some $I_x$ is approximately equal to*

$$(d+1)\cdot\binom{n}{2}^{-e}\sum_{y=2}^{\min(n,2e)}\sum_{i=0}^{y}(-1)^i\binom{n}{y}\binom{y}{i}\binom{y-i}{2}^e\Phi(n-y,n,d,k,t,p).$$

*The expected number k-complexes with intersection equal to some $I_x$ is equal to the number of positive k-complexes identified by our method.*

EXAMPLE 3. Suppose $t = 100$ and there are six positive 4-complexes, i.e., $d = 5$, $k = 4$. Assume first stage error rates of 1% , 3% , and 5% and an error-free confirmatory stage. If we use a $M^2(34,0.77,100)$ pooling design and decoding strategy, then, using PROPOSITION 1, an expected number of at least 3.12, 2.37, 1.42 positive k-complexes are identified respectively using an expected number of 639 tests.

## 8. References

1. Balding, D. J., and Torney, D. C., Optimal pooling designs with error detection, J. Combinatorial Th. A. 74, 1996. 131-141

2. Balding, D. J., et al., A comparative survey of non-adaptive pooling designs, Genetic Mapping and DNA Sequencing, 133-155, IMA Volumes in Mathematics and its Applications, Springer Verlag, 1995.

3. Bruno, W. J., et al., Design of efficient pooling experiments, Genomics, 26, 1995, 21-30.

4. Du, D-Z. and Hwang, F. K. Combinatorial Group Testing and Its Applications, World Scientific, Singapore,1993.

5. Farach, et al. Group testing problems with sequences experimental molecular biology, Proceedings of Compression and Complexity of Sequences, 1997, B. Carpentieri, et al. (Eds.) IEEE Press, 357-367

6. Knill, E. and Muthukrishnan, S.Group testing problems in experimental computational biology, Technical Report LAUR-95-150, Los Alamos National Laboratory,1994.

7. Knill, E., et al. Interpretation of Pooling Experiments Using Markov Chain Monte Carlo Method, Journal of Computational Biology, No. 3, 3, 1996, 395-406

8. Knill, E., et al., Non-adaptive group testing in the presence of errors, Discrete Applied Mathematics, 88, 1998, 261-290

9. Le, D. and Macula, A., On the Probability that Subset Sequences are Minimal, Discrete Mathematics, 207, 1999, 285-289

10. Macula, A., A simple construction of d-disjunct matrices with certain constant weights, Discrete Mathematics, 162, 1996, 311-312

11. Macula, A., Probabilistic Nonadaptive Group Testing in the Presence of Errors and DNA Library Screening, Annals of Combinatorics, 3, 1999, 61-69

12. Macula. A and Rykov, V., Probabilistic Nonadaptive Group Testing In The Presence Of Errors And Inhibitors, J. Stat. Plan. and Inf., to apper

13. Olson, M., et al., A common language for physical mapping of the human genome, Science 245, 1989., 7474-7476

14. Torney, D, Sets Pooling Designs, Annals of Combinatorics,3, 1999, 95-101

DEPARTMENT of MATHEMATICS, SUNY GENESEO, GENESEO, NY 14454
*E-mail adress*: macula@geneseo.edu

THEORETICAL BIOLOGY and BIOPHYSICS, K710, LOS ALAMOS NATIONAL LABORATORY, LOS ALAMOS, NM 87544
*E-mail adress*: dct@ipmati1.lanl.gov

FACULTY of MECHANICS and MATHEMATICS, MOSCOW STATE UNIVERSITY, MOSCOW, RUSSIA
*E-mail adress*: paul@vilenkin.dnttm.ru

DIMACS Series in Discrete Mathematics
and Theoretical Computer Science
Volume **55**, 2000

# On the Perfectness of Tagged Probe Interval Graphs

Li Sheng, Chi Wang, and Peisen Zhang

ABSTRACT. A generalization of interval graph is introduced for cosmid contig mapping of DNA. A graph is a tagged probe interval graph if its vertex set can be partitioned into two subsets of probes and nonprobes, and a closed interval can be assigned to each vertex such that two vertices are adjacent if and only if at least one of them is a probe and one end of its corresponding interval is contained in the interval corresponding to the other vertex. We show that tagged probe interval graphs are weakly triangulated graphs, hence are perfect graphs.

## 1. Cosmid Contig Mapping and Tagged Probe Interval Graphs

A graph $G = (V, E)$ is an *interval graph* if one can map each vertex into an interval on the real line so that two vertices are adjacent if and only if their corresponding interval intersects. The notion of interval graphs was initiated by Seymour Benzer in 1959 [**1**] to study the hypothesis of the linear structure of the fine gene inside the DNA. Interval graphs have since then been widely used in physical mapping of DNA for reconstruction of relative positions of DNA fragments [**12, 13, 16**]. In this paper, we introduce a generalization of interval graph for physical mapping of DNA. Structural properties related to algorithmic development are also studied.

In molecular biology, to replicate and study a certain contiguous stretch of DNA, DNA is cut into small fragments. Those fragments are replicated into clones. The goal of physical mapping is to reconstruct the relative positions of clones along the original DNA. When complete overlap information among the clones is available and correct, there are efficient graph theory algorithms that produce a list of map candidates based on the interval graph model [**2, 8, 7, 10**]. On the other hand, many mapping techniques can only produce partial overlap information.

One mapping technique – cosmid contig mapping, generates overlap information by hybridization [**4, 18, 9, 15**]. In this process, a set of clones is placed on a filter for colony hybridization. The filter is probed with clones that have been radioactively labeled at their ends. We detect that two clones overlap if one of them is labeled and at least one end of this labeled clone is contained in the other clone.

---

2000 *Mathematics Subject Classification.* 05C90, 05C17.

The first author was supported in part by NSF Grant SES-9211492.
Research supported by University of Louisville IRIG grant.

The traditional interval graph model is not applicable for cosmid contig mapping because of the lack of complete overlapping information.

To model cosmid contig mapping and develop mathematical algorithms to construct maps, we introduce a generalization of interval graphs that capture the property of overlap information from cosmid contig mapping. Let $G = (V, E)$ be a graph whose vertex set is $V$ and edge set is $E$ . Let $V = \{P, N\}$ be a partition of $V$. We call $G$ a *tagged probe interval graph with respect to* $P$ if there is a set of closed intervals $I = \{I_v = [l_v, u_v] | v \in V\}$ such that vertices $u$ and $v$ are adjacent in $G$ if and only if one of the following holds:

(1) $u \in P$ and either $l_u \in I_v$ or $r_u \in I_v$, or
(2) $l_v \in I_u$ or $r_v \in I_u$ and $v \in P$

We call vertices in $P$ *probes* and call vertices in $N$ *nonprobes.* $I$ is called a *tagged (probe interval) representation* of $G$ with respect to the given partition. We call a graph $G = (V, E)$ a *tagged probe interval graph* if there is a partition $V = \{P, N\}$ such that $G$ is a tagged probe interval graph with respect to $P$. In practice, a partition, i.e., pre-selected probes and nonprobes, is often given by biologists. When such a partition $V = \{P, N\}$ is given, we denote the graph by $G = (V, P, E)$ to emphasize the given partition.

In cosmid contig mapping, radioactively labeled clones correspond to vertices in $P$ and the other clones correspond to vertices in $N = V - P$. The task in cosmid contig mapping can now be interpreted as testing if a graph with a given partition is a tagged probe interval graph and constructing a tagged representation for $G$ when $G$ is a tagged probe interval graph.

## 2. Preliminaries

Let $G = (V, E)$ be a graph. For a subset $S \subseteq V$, denote $G(S)$ the subgraph of $G$ *induced* by $S$. Denote $N(x) = \{u \in V | ux \in E\}$ the *neighborhood* of $x \in V$ and denote $N[x] = N(x) \cup \{x\}$ the *closed neighborhood of* $x$.

Tagged probe interval graphs introduced here can be viewed as a refinement of probe interval graphs [**18, 17, 9, 15**]. A graph $G = (V, E)$ is a *probe interval graph* (**PIG**) if the vertex set $V$ can be partitioned into subsets $P$ and $N$, and we can assign every vertex $v \in V$ an interval $I_v$ such that $uv \in E$ if and only if $I_u \cap I_v \neq \emptyset$ and $\{u, v\} \cap P \neq \emptyset$, where $I$ is called a *probe (interval) representation of* $G$ with respect to $P$. Probe interval graphs were also introduced for cosmid contig mapping. If we assume that probes are not only labeled on their two ends, but also on their interior points, then probe interval graphs can be used to represent the overlap information obtained.

The difference between an probe interval graph and an tagged probe interval graph is that for a probe interval graph $G = (V, P, E)$, and for any two vertices $u \in N, \in P$, $uv$ is an edge as long as $I_u$ and $I_v$ have a non empty intersection. Whereas for a tagged probe interval graph with $u \in N, v \in P$, even if $I_u \cap I_v \neq \emptyset$, $uv$ may not be an edge unless that $I_u$ is not completely contained in $I_v$. In figure 1, we consider the graph $G = C_4$, with three darken vertices $a, b, d$ being probes. $G$ is a tagged probe interval graph with respect to the give partition, and its tagged representation is given in the Figure 1. Notice that although the intervals for $a, c$ intersect, $ac$ is not an edge because the interval $I_c$ for the nonprobe vertex $c$ is completely contained in the interval $I_a$. But $G$ is not a probe interval graph with repect to the given partition, because one can show that no matter how you layout the interval, since the two probes $b, d$ do not intersect, the intervals for the other two vertices $a, c$ must intersect, and thus, $ac$ must be an edge in $G$ which reaches a contradiction.

FIGURE 1. A TPIG but not PIG with respect to the given partition.

In Sheng [**14**], examples are also given for graphs that are probe interval graphs with respect to some partitions but tagged probe interval graphs with respect to any partition,, and vice versa. That is there is no containment relation between tagged probe interval graphs and probe interval graphs.

An interval graph is both a tagged probe interval graph and a probe interval graph with respect to $P = V$. A tagged probe interval graph or a probe interval graph may not be an interval graph. For example, $C_4$ is not an interval graph, but one can easily find a partition of $P$ and $N$ on its vertex set to make $C_4$ a (tagged) probe interval graph.

The following lemma summarizes some observations which will be used in the remainder of this paper.

LEMMA 2.1. *Suppose $G = (V, P, E)$ is a tagged probe interval graph and $I = \{I_v | v \in V\}$ is a representation of $G$ with respect to the given partition.*

(1) *$G(P)$ is an interval graph. In particular, if $P = V$ then $G$ is an interval graph;*
(2) *$G(V - P)$ is an independent set;*
(3) *If $H \subseteq V$, then $G(H)$ is also a tagged probe interval graph;*
(4) *If $u \in P$, $v \in N$ and $uv \notin E$, then $I_v$ not properly contained in $I_u$ implies that $I_u \cap I_v = \emptyset$;*
(5) *If $|\{u, v\} \cap P| = 1$ and $uv \notin E$ imply $I_u \cap I_v = \emptyset$ for any $u, v \in V$, then $G$ is a probe interval graph with respect to $P$ and $I$ is a probe representation of $G$ with respect to $P$.*

**Proof.** (1) - (4) follow directly from the definition. Condition (5) implies that $uv \in E$ if and only if at least one of $u, v$ is a probe and $I_u \cap I_v \neq \emptyset$. Hence $G$ is a probe interval graph and the lemma follows. □

## 3. Perfectness of Tagged Probe Interval Graphs

In this section, we show that tagged probe interval graphs are perfect graphs. Given graph $G = (V, E)$, the *clique number* $\omega(G)$ is the number of vertices in a maximum complete subgraph of $G$. The *chromatic number* $\chi(G)$ *of* $G$ is the minimum number of independent sets needed to cover all vertices of $G$. $G$ is *perfect* if $\omega(H) = \chi(H)$ for all induced subgraphs $H$ of $G$.

Perfect graphs are very important in combinatorial optimization and algorithmic theory. Interval graphs are well known perfect graphs. Many classic graph theoretic problems (such as coloring and hamiltonian path problems) that are NP-complete in general are found to be polynomial solvable for perfect graphs [**3, 11**].

Because of the large amount and variety of the data from cosmid contig mapping, efficient algorithms are essential. Thus, the perfectness on the (tagged) probe interval graph not only expands the classes of perfect graph family, but also enables the development of fast algorithms to calculate graph parameters and to reconstruct, combine and refine physical maps of DNA.

A graph is *weakly triangulated* if both $G$ and $G^c$, the complement of $G$, do not have any induced subgraph isomorphic to a chordless cycle of at least five vertices. Hayword [**6**] showed that weakly triangulated graphs are perfect graphs. The following was shown in [**9**].

THEOREM 3.1. (McMorris, Wang and Zhang [**9**]) *Probe interval graphs are weakly triangulated.*

In the following, we show that, similar to probe interval graphs, tagged probe interval graphs are weakly triangulated graphs.

LEMMA 3.2. *A chordless cycle $C_k$ $(k \geq 5)$ is not a tagged probe interval graph.*

**Proof.** Let $C_k = (V, E)$ $(k \geq 5)$ be a chordless cycle. Suppose on the contrary that $\{I_v | v \in V\}$ is a tagged representation of $C_k$ $(k \geq 5)$ with respect to some $P \subseteq V$. Then $V \neq P$ since $C_k$ is not an interval graph. Suppose $u, v$ are a pair of nonadjacent vertices such that $u \in P$ and $v \in N$. By (2) of Lemma 2.1, the two vertices adjacent to $v$ in $C_k$, say $x$ and $y$, must be probes. Since $k \geq 5$ and $uv \notin E$, it follows that $\{x, y\} \nsubseteq N[u]$ and $u \neq x, y$. By (1) of Lemma 2.1, either $I_u \cap I_x = \emptyset$ or $I_u \cap I_y = \emptyset$. Therefore $I_v$ cannot be properly contained in $I_u$. Then $uv \notin E$ implies that $I_u \cap I_v = \emptyset$. By (5) of Lemma 2.1 $C_k$ would be a probe interval graph, contradicting Theorem 3.1. □

LEMMA 3.3. *The complement of any chordless cycle $C_k$ $(k \geq 5)$ is not a tagged probe interval graph.*

**Proof.** Let $C_k^c = (V, E)$ be the complement of $C_k$. Since $C_5^c = C_5$, $C_5^c$ is not a tagged probe interval graph with respect to any $P \subseteq V$ by Lemma 3.2. For $k \geq 6$, assume on the contrary that $I = \{I_v | v \in V\}$ is a tagged representation of $C_k^c$ with respect to some $P \subseteq V$. Since $C_k^c$ has induced subgraphs isomorphic to $C_4$, by (1) of Lemma 2.1, $P \neq V$. Similar to the proof of Lemma 3.2, let $u, v$ be a pair of nonadjacent vertices of $C_k^c$ such that $u \in P$ and $v \in N$. In the representation $I_v$ cannot be properly contained in $I_u$ for otherwise any vertex adjacent to $v$ is a probe and hence adjacent to $u$ by (1) of Lemma 2.1. It follows from $uv \notin E$ that $I_u \cap I_v = \emptyset$. By (5) of Lemma 2.1, $C_k^c$ $(k \geq 6)$ would be a probe interval graph, contradicting Theorem 3.1. □

THEOREM 3.4. *Tagged probe interval graphs are weakly triangulated.*

**Proof.** By (3) of Lemma 2.1 an induced subgraph of a tagged probe interval graph is also a tagged probe interval graph. Therefore by Lemmas 3.2 and 3.3, a tagged probe interval graph cannot contain any induced subgraph isomorphic to $C_k$ or $C_k^c$ $(k \geq 5)$. □

COROLLARY 3.5. *Tagged probe interval graphs are perfect.*

## References

[1] S. Benzer. On the topology of genetic fine structure. In *Proc. Nat. Acad. Sci.*, volume **45**, pages 1607–1620, USA, 1959.

[2] K.S. Booth and S.G. Lueker. Testing for consecutive ones property, interval graphs and planarity using PQ-tree algorithms, *J. Comput. Syst. Sci.* 13(1976), 335—379.

[3] J.S. Deogun and G. Steiner. Hamiltonian cycle is polynomial on cocomparability graphs. *Discrete Appl. Math.*, **39**:165–172, 1993.

[4] S.G. Fischer, E. Cayanis, J. Russo, I. Sunjevaric, B. Boukhgalter, X.L. Li, P. Zhang, P. Rothstein, M.T. Yu, D. Warburton, I.S. Edelman and A. Efstratiadis. Assembly of ordered contigs from YAC-selected cosmids of human chromosome 13, *Genomics* 21 (1994), 525-537.

[5] M.C. Golumbic. *Algorithmic Graph Theory and Perfect Graphs.* Academic Press, New York, 1980.

[6] R.B. Hayward. Weakly triangulated graphs, *J. Comb. Theory (B)* 39 (1985), 200—209.

[7] W. L. Hsu. A simple test for the consecutive ones property.' In *Proceedings of the International Symposium on Algorithms and computation*, volume **650** of *Lecture Notes in Comput. Sci.*, pages 459–468. Springer, Berlin, 1992.

[8] N. Korte and R. H. Möhring. An incremental linear-time algorithm for recognizing interval graphs. *SIAM J. Comput.*, **18**:68–81, 1989.

[9] F.R. McMorris; Chi Wang; Peisen Zhang, On probe interval graphs. Discrete Appl. Math. 88 (1998), no. 1-3, 315–324.

[10] J. Meidanis and E. G. Munuera. A theory for the consecutive ones property. In *Proceedings of the Third South American Workshop on String Processing*, volume **4** of *International Information Series*, pages 194–202. Carleton University Press, 1996.

[11] G. Narasimhan and R. Manber. Stability and chromatic number of tolerance graphs. *Disc. App. Math.*, **36**:47–56, 1992.

[12] F. S. Roberts. *Discrete Mathematical Models, with Applications to Social, Biological, and Environmental Problems.* Prentice-Hall, Englewood Cliffs, N.J., 1976.

[13] F. S. Roberts. Seven fundamental ideas in the application of combinatorics and graph theory in biological and social sciences. In F. S. Roberts, editor, *Applications of Combinatorics and Graph Theory in the Biological and Social Sciences*, volume **17** of *IMA Volumes in Mathematics and its Applications*, pages 1–37. Springer-Verlag, New York, 1989.

[14] L. Sheng. *Some Graph Theoretic Approaches to Problems of the Social and Biological Sciences: Social Roles, Phylogenetic Trees, and Physical Mapping.* PhD thesis, RUTCOR–Rutgers Center for Operations Research, Rutgers University, New Brunswick, New Jersey 08903, 1998.

[15] L. Sheng. Cycle Free Probe Interval Graphs, *Congressus Numerantium* 1999, in Press.

[16] M. S. Waterman and J. R. Griggs. Interval graphs and maps of DNA. *Bull. Math. Biol.*, **48**:189–195, 1986.

[17] P. Zhang. Probe interval graphs and its applications to physical mapping of DNA, *manuscript*, 1994.

[18] P. Zhang, E.A. Schon, S.F. Fischer, E. Cayanis, J. Weiss, S. Kistler and P.E. Bourne. An algorithm based on graph theory for the assembly of contigs in physical mapping of DNA, *CABIOS* 10 (1994), 309-317.

Department of Mathematics and Computer Science, Drexel University, Philadelphia, PA 19104

*Current address*: Department of Mathematics and Computer Science, Drexel University, Philadelphia, PA 19104

*E-mail address*: `lsheng@mcs.drexel.edu`

Department of Mathematics, University of Louisville, Louisville, KY 40290

*Current address*: Department of Mathematics, University of Louisville, Louisville, KY 40290

*E-mail address*: `chiwang@louisville.edu`

Department of Genetics and Development, Columbia University, New York, NY 10032

*Current address*: Department of Genetics and Development, Columbia University, New York, NY 10032

*E-mail address*: `peisen_z@yahoo.com`

DIMACS Series in Discrete Mathematics
and Theoretical Computer Science
Volume **55**, 2000

# Frequent Sets in the Presence of Clones: An Example from Medical Surveillance

Stephen E. Brossette and Alan P. Sprague

ABSTRACT. An important, but little noted challenge for hospitals is to control outbreaks of infectious disease, especially those caused by bacteria resistant to multiple antibiotics. To address this challenge, we use Data Mining techniques to enhance hospital infection control surveillance, and are developing new frequent set algorithms, more efficient than traditional ones, to compute frequent sets for data that describe multiple isolates of highly resistant clones of bacteria. These algorithms use an equivalence relation on frequent sets.

## 1. Introduction

An algorithmic and combinatorics problem was found in the application of data mining techniques to hospital infection control surveillance. The difficulty arose from a combinatorial explosion in the lattice of "frequent" subsets of a set I, and can be resolved by recognizing an equivalence relation on the members of the lattice, and generating at least one but preferably few members of each equivalence class. We start with a brief description of hospital infection control surveillance.

A subset of hospital infection control surveillance is performed in the following manner. First, samples of sputum, blood, wound tissue, etc. are taken from infected hospital patients. Each sample is cultured and the bacteria isolated are tested against an array of antibiotics. Results are entered into a database. Each record in the database describes a single bacterial isolate, and contains patient information such as sex, race, and location in the hospital. This data is reviewed for aberrancies by hospital infection control officers. Using traditional methods, if a problem is not suspected it will not likely be detected. In contrast, our Data Mining techniques enable us to identify unexpected and potentially interesting patterns.

In our system, data is separated into one month partitions, and at the end of each month that month's data is compared to the previous several months [**6, 5**]. We are interested in discovering patterns such as:

(1) Of the 80 isolates of Pseudomonas this month, 28 were resistant to both antibiotics $a_1$ and $a_2$, which is substantially higher than the incidence of resistance in previous months.
(2) Of the 84 isolates from the intensive care unit, 4, each a different organism, were resistant to antibiotics 1 through 5. This pattern of resistance in any organism had not previously been seen.

2000 *Mathematics Subject Classification.* 06D04,68R10.

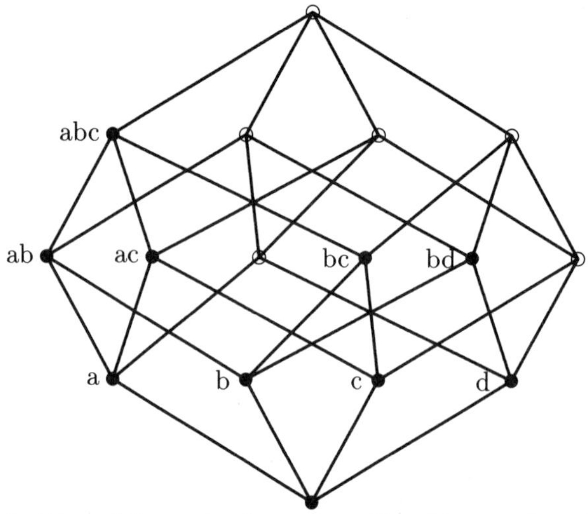


FIGURE 1. Lattice of itemsets, for a data file of 5 records, where $I = \{a, b, c, d\}$. The 5 records are $\{a, b, c\}$, $\{a, b, c\}$, $\{a, b, d\}$, $\{b, c, d\}$, $\{a, c\}$. Where the frequency threshold $\theta$ is 2, solid itemsets are frequent and hollow ones are not.

The first step in the processing on each month's data is the generation of frequent sets. In Section 2 we describe frequent sets and an algorithm to generate frequent sets. The combinatorial explosion that arose while generating frequent sets on a data file is described in Section 3. Section 4 describes progress toward a new algorithm for generating frequent sets. Section 5 concludes, emphasizing the principal contribution of this paper to Data Mining.

## 2. Frequent Sets

In this section we define frequent sets, make some simple observations about them, and describe a well-known algorithm to generate them.

Let $I$ be a set, whose members will be called "items". Let $D$ be a file, i.e. a set of records. Each record is a subset of $I$. (Several records may correspond to the same subset of $I$.) For example, in our application of hospital infection control surveillance, $I$ may be { Male, Female, ICU, ..., Pseudomonas, ..., $a_1, a_2$, ... }, where $a_i$ means "resistant to the $i$-th antibiotic". In our application a record in $D$ is the set of items pertaining to a single isolate.

Any subset of $I$ will be called an *itemset*. For an itemset $J$, we let $R(J)$ be the set of records containing $J$. The *support* of $J$ (written $\text{suppt}(J)$) equals $|R(J)|$; i.e., it is the number of records $r$ such that every item in $J$ appears in $r$. $J$ is a *frequent set*, or simply *frequent*, if $\text{suppt}(J) \geq \theta$, where $\theta$ is some preset threshold.

We note that if itemsets $J_1$ and $J_2$ satisfy $J_1 \subseteq J_2$ then $\text{suppt}(J_1) \geq \text{suppt}(J_2)$. Consequently, if $J_1 \subseteq J_2$ and $J_2$ is frequent then $J_1$ is also frequent.

Figure 1 shows the lattice of itemsets for a data file of 5 records, where $|I| = 4$.

Various algorithms to generate frequent sets are known ([**1, 4, 7**]). We give high-level pseudocode for an algorithm to generate all frequent sets. This algorithm is called Apriori, and probably is the most commonly cited algorithm in the literature ([**1, 2**]).

Apriori

1. $F_1$ = all frequent 1-itemsets.
2. $k = 2$.
3. While $F_{k-1}$ is nonempty:
4.     /* Using $F_{k-1}$, generate $k$-item frequent sets $F_k$. */
5.     $C_k$ = all $k$-itemsets $J$ such that all $k$ $(k-1)$-item subsets of $J$ are in $F_{k-1}$.
6.         For all $J$ in $C_k$:
7.             Compute the support of $J$.
8.             If $\text{suppt}(J) \geq \theta$:
9.                 Put $J$ into $F_k$.

In this algorithm, $C_k$ is the set of *candidates*: every $k$-item frequent set (as well as many which are not frequent) is placed into $C_k$. Those members of $C_k$ having large support are indeed frequent, and are placed into $F_k$ (the set of $k$-item frequent sets).

## 3. Combinatorial Explosion

We have applied our system to 15 months of data from UAB Hospital (1997 through early 1998). We also chose to test our system on a known outbreak of highly resistant bacteria, Acinetobacter baumannii, which occurred in 1994. The 1994 data no longer exists, but since it was described in detail we were able to reproduce it in 1997 data. When our system was given this modified data, it crashed secondary to insufficient memory while generating frequent sets. The host computer had what was by 1998 standards a reasonable amount of memory, and the program was designed more to execute rapidly than to conserve space. The data files were not especially large — 140 items, 500 records, an average record containing 6.6 items.

So why was so much space needed? If one bacterial clone is responsible for a cluster of several isolates, each isolate in the cluster will test resistant to the same antibiotics, and each corresponding record will contain identical lists of antibiotic names. If such a clonal cluster is resistant to many antibiotics, an occurrence that is unfortunately becoming more common, then each record in the cluster will contain the same long list of antibiotic names. This antibiotic name list alone can be 10 to 20 items long. If the clonal cluster is limited to a specific location or patient population, the number of identical items in the corresponding records grows even further. The data file that contained the Acinetobacter baumannii clonal cluster contained a frequent set of 17 items. This immediately implies that the data file gave rise to at least $2^{17}$ frequent sets.

Here is a short illustrative example. The data set shown here contains only 4 records:

1. Org1, Loc1, Antb1, Antb3.
2. Org1, Loc2, Antb1, Antb2, ..., Antb12.
3. Org1, Loc2, Antb1, Antb2, ..., Antb12.
4. Org2, Loc2, Antb3, Antb4.

Records 2 and 3 describe a clonal cluster of Organism1 in Location2 that is resistant to antibiotics 1 through 12. Record 1 describes an isolate of Organism1 that is not part of the clonal cluster just described. Record 4 describes an isolate of Organism2.

We want to generate frequent sets from this data. We say that two itemsets $J_1$ and $J_2$ are *equivalent* if $R(J_1) = R(J_2)$, i.e. a record contains $J_1$ iff it contains $J_2$. This is an equivalence relation.

With a frequent set support threshold of 2, the example yields many equivalent frequent sets. For example, {Org1, Loc2} and {Org1, Loc2, Antb2} are equivalent because all records that contain Org1 and Loc2 also contain Antb2. With the Apriori algorithm, 16384 frequent sets are generated, all belonging to only 4 equivalence classes. Four inequivalent frequent sets are {Antb3}, {Loc2, Antb3, Antb4}, {Org1, Antb1, Antb3}, and {Org1, Loc2, Antb1, Antb2, ... , Antb12}. Their supports are, respectively, 4, 3, 3, and 2.

In the next section, we describe an algorithm that generates only 19 frequent sets instead of 16384. On infection control data containing an Acinetobacter baumannii clonal cluster, the new algorithm generates 2401 frequent sets, compared to over 150,000 by Apriori.

## 4. Algorithms for Highly Clonal Data

In this section we present an algorithm that is better suited for highly clonal data, and describe ongoing work. In the previous section, we defined two itemsets $J_1$ and $J_2$ to be equivalent if $R(J_1) = R(J_2)$. Where $J_1$ and $J_2$ are equivalent, we write $J_1 \equiv J_2$. An equivalence class by this relation will be called a *clone*.

We observe that if $J_1 \equiv J_2$ then $J_1 \equiv J_1 \cup J_2$. Hence each clone has exactly one maximal itemset, namely the union of all itemsets in the clone.

For each itemset $J$, we define the closure of $J$ (written $\overline{J}$) to be the maximum itemset in the clone of $J$. Equivalently, $\overline{J} = \cap\{r : r \in R(J)\}$.

Apriori and other traditional frequent set algorithms generate all frequent sets. As we noted in the previous section, this can be a very resource-consuming objective. When the data contain large clones, little information is lost by generating only a few frequent sets of each clone. There are two contrasting goals that one might adopt when generating representative members of each clone.

- For each clone, generate all frequent set which are *minimal* in their clone.
- For each clone, generate all frequent set which are *maximal* in their clone.

Bayardo discusses an algorithm for the first goal in [**3**]. For our purposes, the second goal is the one of interest.

To implement this, we employ a 2-phase process. First, generate frequent sets that are minimal in their clone. Second, compute the closure of each frequent set generated.

For an itemset $J$, each itemset $J - \{a\}$ where $a \in J$ will be called a *parent* of $J$. Where $J$ is a frequent set of $k$ items, $J$ has $k$ parents, which are also clearly frequent. Where $J$ is minimal in its clone, it may be shown that each of its parents $J - \{a\}$ is minimal in its clone (that is, minimal in the clone to which $J - \{a\}$ belongs, which evidently is different from the clone to which $J$ belongs, by the minimality of $J$).

The clone algorithm has two phases. In the first phase, all frequent sets which are minimal in their clone are generated. Pseudocode for Phase 1 is the same as the above pseudocode for Apriori, but with line 9 in Apriori replaced by:

```
9.            If suppt(J) is less than the support of each of its parents:
10.               Put J into F_k.
```

For Phase 2, there are two straightforward ways to compute the closure of each frequent set. One method is, for each frequent set $J$ generated in Phase 1, to compute $\overline{J} = \cap\{r : r \in R(J)\}$. An alternate method is to compute $\overline{J}$ as $J \cup \{a \in I - J : R(a) \supseteq R(J)\}$. Both of these methods are somewhat computation intensive.

We are developing algorithms to decrease the computation load in Phase 2. The general idea is to use information gleaned in Phase 1 to decrease the amount of computation in Phase 2. We are currently focusing our attention on one such algorithm, which has an interesting proof of correctness.

## 5. Conclusion

We feel that our principal contribution to Data Mining is to notice and interpret an as-yet little noted computational burden in the generation of frequent sets. While there is an extensive literature describing algorithms for generating frequent sets [**1, 2, 3, 4, 7**], many focus only on the challenges posed by massive input files. Only Bayardo has recognized the redundant computation by Apriori.

While Bayardo chooses to generate frequent sets minimal in their clone, we generate those that are maximal, the harder of the two goals. We have described a method that first generates frequent sets minimal in their clone, then infers those that are maximal. Accomplishing the second phase with as little computation as possible yields a new algorithm with an interesting proof which will be described in an upcoming manuscript.

## References

1. R. Agrawal, H. Mannila, R. Srikant, H. Toivonen, and A.I. Verkamo, *Fast discovery of association rules*, Advances in Knowledge Discovery and Data Mining (U. Fayyad, G. Piatetsky-Shapiro, P. Smyth, and R. Uthurusamy, eds.), AAAI Press, 1996, pp. 307–328.
2. R. Agrawal and R. Srikant, *Fast algorithms for mining association rules*, Proc. 20th VLDB Conf., 1994.
3. R.J. Bayardo, *Brute-force mining of high-confidence classification rules*, Proc. Third Intl. Conf. Knowledge Discovery Data Mining, 1997, pp. 123–126.
4. S. Brin, R. Motwani, J.D. Ullman, and S. Tsur, *Dynamic itemset counting and implication rules for market basket data*, Proc. ACM SIGMOD Conf. Mgmt. Data, 1997, pp. 255–264.
5. S.E. Brossette, W.T. Jones, A.P. Sprague, and S.A. Moser, *Knowledge discovery system applied to hospital epidemiology*, submitted.
6. S.E. Brossette, A.P. Sprague, J.M. Hardin, K.B. Waites, W.T. Jones, and S.A. Moser, *Association rules and data mining in hospital infection control and public health surveillance*, J. Amer. Medical Informatics Assn. **5** (1998), 373–381.
7. A. Savasere, E. Omiecinski, and S. Navathe, *An efficient algorithm for mining association rules in large databases*, Proc. Intl. Conf. Very Large Data Bases, 1995, pp. 432–444.

Department of Computer and Information Sciences, University of Alabama at Birmingham, Birmingham, Alabama 35294-1170

*E-mail address*: sbrossette@mindspring.com

Department of Computer and Information Sciences, University of Alabama at Birmingham, Birmingham, Alabama 35294-1170

*E-mail address*: sprague@cis.uab.edu

DIMACS Series in Discrete Mathematics
and Theoretical Computer Science
Volume **55**, 2000

# A Survey on Combinatorial Group Testing Algorithms with Applications to DNA Library Screening

Hung Q. Ngo and Ding-Zhu Du

ABSTRACT. In this paper, we give an overview of Combinatorial Group Testing algorithms which are applicable to DNA Library Screening. Our survey focuses on several classes of constructions not discussed in previous surveys, provides a general view on pooling design constructions and poses several open questions arising from this view.

## 1. Introduction

The basic problem of DNA library screening is to determine which *clone* (a DNA segment) from the library contains which *probe* from a given collection of probes in an efficient fashion. A clone is said to be *positive* for a probe if it contains the probe, and *negative* otherwise. In practice clones are pooled together in some manner to be tested against each probe, since checking each clone-probe pair is expensive and usually only a few clones contain any given probe. An example is when Sequenced-Tagged Site markers (also called STS probes) are used [**OHCB89**]. If the test result for a pool (of clones) is negative, indicating that no clone in the pool contains the probe, then no further tests are needed for the clones in the pool.

This problem is just an instance of the general group testing problem, in which a large population of *items* containing a small set of *defectives* are to be tested to identify the defectives efficiently. We assume some testing mechanism exists which if applied to an arbitrary subset of the population gives a *negative outcome* if the subset contains no defective and *positive outcome* otherwise. Objectives of group testing vary from minimizing the number of tests, limiting number of pools, limiting pool sizes to tolerating a few errors. It is conceivable that these objectives are often contradicting, thus testing strategies are application dependent.

Group testing algorithms can roughly be divided into two categories : *Combinatorial Group Testing* (CGT) and *Probabilistic Group Testing* (PGT). In CGT,

---

2000 *Mathematics Subject Classification.* Primary 05B20, 05D05; Secondary 05D40, 51E30.

Support in part by by the National Science Foundation under grant CCR-9530306.

it is often assumed that the number of defectives among $n$ items is equal to or at most $d$ for some fixed positive integer $d$. In PGT, we fix some probability $p$ of having a defective. If the pools are simultaneously tested $s$ times, with later test pools collected based on previous test results, then the CGT algorithm is said to be an $s$-stage algorithm. Group testing strategies can also be either *adaptive* or *non-adaptive*. A group testing algorithm is non-adaptive if all tests must be specified without knowing the outcomes of other tests. Clearly, being non-adaptive is equivalent to being 1-stage. A group testing algorithm is *error tolerant* if it can detect or correct some $e$ errors in test outcomes.

Library screening applications introduce several new constraints to group testing. Firstly, $s$-stage group testing algorithms with small $s$ (e.g. $\leq 2$) are often preferable [**BT96, BBKT96**]. The common requirement is to have an adaptive algorithm. Secondly, DNA screening is error prone since the pools have to be purified before probing. Hence, tolerating several errors is desirable [**BT96**]. Lastly, as assembling pools is costly, sometime robots are used to assemble the pools. This makes coordinating the pools with some physical arrangement of clones (such as a grid) important.

As far as we know, there are three related surveys previously done in this area. The first was a survey from Dyachkov and Rykov (1983, [**DR83**]) done in the context of superimposed codes. The second was a monograph by Du and Hwang (1993, [**DH93**]), which gave a nice account of CGT algorithms. The third was an article by Balding et al. (1995, [**BBKT96**]), which comparatively surveyed certain classes of non-adaptive algorithms.

In this paper, we give an overview of Combinatorial Group Testing algorithms with applications to DNA Library Screening. Our survey focuses on several classes of constructions not discussed in previous surveys, provides a general view on pooling design constructions and poses several open questions arising from this view.

The rest of the paper is organized as follows. Section 2 fixes up basic definitions and notations needed for the rest of the paper. It also gives a taxonomy of non-adaptive group testing algorithms from which later sections are organized. Section 3 discusses deterministic algorithms. Section 4 provides a new general perspective on constructing a class of deterministic pooling designs, from which several open problems popped up naturally. Section 5 presents random algorithms, and section 6 introduces error-tolerance group testing algorithms. Section 7 concludes the paper.

## 2. Preliminaries

**2.1. The Matrix Representation.** We first emphasize that we are concerned only with combinatorially non-adaptive group testing strategies, for DNA library screening applications prefer parallel tests as we have mentioned earlier. The "combinatorial" part comes from the assumption that there are at most $d$ defectives in a population of $n$ items.

Consider a $v \times n$ 01-matrix $M$. Let $R_i$ and $C_j$ denote row $i$ and column $j$ respectively. Abusing notation, we also let $R_i$ (resp. $C_j$) denote the set of column (resp. row) indices corresponding to the 1-entries of row $i$ (resp. column $j$). The *weight* of a row or a column is the number of 1's it has. $M$ is said to be *d-disjunct* if the union of any $d$ columns does not contain another. A $d$-disjunct $v \times n$ matrix $M$ can be used to design a non-adaptive group testing algorithm on $n$ items by associating the columns with the items and the rows with the pools to be tested. If $M_{ij} = 1$ then item $j$ is contained in pool $i$ (and thus test $i$). If there are no more than $d$ defectives and the test outcomes are error-free, then it is easy to see that the test outcomes uniquely identify the set of defectives. We simply identify the items contained in negative pools as *negatives* (good items) and the rest as *positives* (defected items). Notice that $d$-disjunct property implies that each set of $\leq d$ defectives corresponds uniquely to a test outcome vector, thus decoding test outcomes involves only a table lookup. The design of a $d$-disjunct matrix is thus also naturally called *non-adaptive pooling design.* We shall use this term interchangeably with the long phrase "non-adaptive combinatorial group testing algorithm".

Let $S(\bar{d}, n)$ denotes the set of all subsets of $n$ items (or columns) with size at most $d$, called the set of *samples*. For $s \in S(\bar{d}, n)$, let $P(s)$ denote the union of all columns corresponding to $s$. A pooling design is $e$-error-detecting (correcting) if it can detect (correct) up to $e$ errors in test outcomes. In other words, if a design is $e$-error-detecting then the test outcome vectors form a $v$-dimensional binary code with minimum Hamming distance at least $e + 1$. Similarly, if a design is $e$-error-correcting then the test outcome vectors form a $v$-dimensional binary code with minimum Hamming distance at least $2e + 1$. The following remarks are simple to see, however useful later on.

REMARK 1. *Suppose $M$ has the property that for any $s, s' \in S(\bar{d}, n), s \neq s'$, $P(s)$ and $P(s')$ viewed as vectors have Hamming distance $\geq k$. In other words, $|P(s) \oplus P(s')| \geq k$ where $\oplus$ denotes the symmetric difference. Then, $M$ is $(k-1)$-error-detecting and $\lfloor \frac{k-1}{2} \rfloor$-error-correcting.*

REMARK 2. *$M$ being d-disjunct is equivalent to the fact that for any set of $d+1$ distinct columns $C_{j_0}, \ldots C_{j_d}$ with one column (say $C_{j_0}$) designated, $C_{j_0}$ has a $1$ in some row where all $C_{j_k}$'s, $1 \leq k \leq d$ contain $0$'s.*

An important question to ask is "given $n$ items with at most $d$ defectives, at least how many tests are needed to identify the defectives?" The best asymptotic answer to this question is dated back to Dyachkov and Rykov (1982, [**DR82**]) and Dyachkov, Rykov and Rashad (1989, [**DRR89**]), which can be summarized by the following theorem.

THEOREM 1. *Let $v(d, n)$ denote the minimum number of pools needed for the $S(\bar{d}, n)$ problem, then as $n \to \infty$ and $d \to \infty$*

$$\frac{d^2}{2\log_2 d}(1 + o(1)) \log_2 n \leq v(d, n) \leq d^2 \log_2 e(1 + o(1)) \log_2 n$$

**2.2. A Taxonomy of Non-Adaptive Pooling Designs.** We now give a tentative taxonomy of non-adaptive pooling designs, from which later sections are organized.

(1) *Deterministic Designs.* This refers to the fact that every pool is deterministically specified. These designs can be further categorized into:
   (i) Set-packing designs.
   (ii) Transversal designs.
   (iii) Designs whose $d$-disjunct matrices are directly constructed.
(2) *Random Designs.* In these designs, some or all of the entries are randomly determined with parameterized probabilities, which could be optimized based on certain objective function(s). The categories are :
   (i) Random matrices.
   (ii) Random weight-$w$ designs.
   (iii) Random size-$k$ designs.
   (iv) Random designs which come from deterministic designs.
(3) *Error Tolerance Designs.* Although these designs are either deterministic or random, they are worth being paid special attention to.

## 3. Deterministic Pooling Designs

**3.1. Set Packing Designs.** First noted by Kautz and Singleton [**KS64**] back in 1964, packing designs with certain parameters can be used to construct disjunct matrices. We first give some basic definitions. A $t$-$(v, k, \lambda)$ *packing design* is a collection $\mathcal{F}$ of $k$-subsets (called *blocks*) of $[v] := \{1, 2 \dots, v\}$ such that any $t$-subset of $[v]$ is contained in at most $\lambda$ members of $\mathcal{F}$. One useful situation for us is when $\lambda = 1$, in which case the packing is called a $(v, k, t)$-packing. Notice that $\lambda = 1$ means no two members of $\mathcal{F}$ have $t$ elements in common. Thus, by Remark 2 if $k > d(t-1)$ a $d$-disjunct matrix $M$ can be constructed from a $(v, k, t)$-packing by simply indexing $M$'s columns by the blocks and $M$'s rows by members of $[v]$. Moreover, by Remark 1 we see that if $k = d(t-1) + q + 1$ $(q \geq 0)$ then $M$ is $q$-error detecting and $\lfloor \frac{q}{2} \rfloor$-error correcting.

Naturally, the basic problem of packing design is to find the *packing number* $D_\lambda(v, k, t)$, the size of a maximum $t$-$(v, k, \lambda)$ packing design. We write $D(v, k, t)$ instead of $D_1(v, k, t)$ when $\lambda = 1$. Maximum sized $(v, k, t)$-packings induce very good pooling designs [**BBKT96**]. Unfortunately, very little is known about optimal packing designs. Most of what we know are for small values of $k$ and $t$. Mills and Mullin [**MM92**] gave a nice account on packing designs. To give the reader a sense of how difficult this problem is, we quote a result on $D(v, k, t)$ in Theorem 2. From the theorem, it is conceivable that finding optimal set packing is just as hard as *the main coding theory problem* [**Rom92**].

THEOREM 2. *Let $A(n, d, w)$ denote the size of a maximum constant $w$-weight binary $(n, d)$-code, then*

$$D(v, k, t) = A(v, 2k - 2t + 2, k)$$

Let

$$U_\lambda(v,k,t) = \left\lfloor \frac{v}{k} \left\lfloor \frac{v-1}{k-1} \cdots \left\lfloor \frac{v-t+1}{k-t+1}\lambda \right\rfloor \right\rfloor \right\rfloor$$

then Schőnheim [**Sch66**] observed that $D_\lambda(v,k,t) \le U_\lambda(v,k,t)$. Equality holds when the design is any $t$-$(v,k,\lambda)$ design. In particular, since we want $\lambda = 1$, Steiner Triple Systems (2-$(v,3,1)$ designs) and Steiner Quadruple Systems (3-$(v,4,1)$ designs) could be used to construct disjunct matrices with small $d$'s. Finite projective planes and affine planes are also $t$-designs with $\lambda = 1$ but they don't give good pooling designs (too many tests). The only other noticeable result which concerns us is from Brouwer [**Bro79**], who determines all values of $D(v,4,2)$. For a comprehensive treatment on design theory, the reader is referred to a nice book by Beth, Jungnickel and Lenz [**BJL86**].

**3.2. Transversal Designs.** The simplest form of transversal designs is called the *grid design*. To facilitate the use of robots for pool assembling, the clones can be arranged into rows and columns of a set of $r \times c$ grids, where each row and column contributes a pool. For simplicity, we can assume $rc \mid n$. Clearly, ambiguity can occur if there are more than one positive clone. The simplest example is when there are two positives, say $a$ and $b$, lying on different rows and columns of a grid $G$. In this case, testing $G$ alone is not enough to identify $a$ and $b$ because the two clones $c$ and $d$ collinear with both $a$ and $b$ are also candidates. To resolve ambiguity, we wish to rearrange $G$ into another grid (giving additional pools) so that $c$ and $d$ are not collinear with both $a$ and $b$ anymore. More grids are needed if there are 3 or more positive clones. In fact, if we require a stronger condition that no two clones are collinear twice, called the *unique collinearity condition*, then Hwang [**Hwa95**] showed that the existence of the grids is equivalent to the existence of certain set of mutually orthogonal Latin squares.

Barillot et al. [**BLC91**] generalized this idea to $k$-dimensional grids, where each intersection point could be viewed as a vertex of the $k$-cube. A new grid $G'$ can be obtained from the old grid $G$ by a linear transformation represented by a matrix $A_{d\times d}$. Thus a vertex $x = (x_1, \dots x_d)^T$ of $G$ is mapped to vertex $Ax$ of $G'$. A third grid could either be obtained by using $A$ twice (with transformation matrix $A^2$) or by using a different transformation matrix $B$ (with transformation matrix $AB$). They also extended the 2-dimensional grid to higher dimension. The set of hyperplanes could be taken as pools, however the pool size is usually large. Reducing pool size by taking lower dimension is possible but that increases the number of tests. Pros and cons of this approach have not been studied.

The general case of *transversal design* was mentioned by Balding et al. in [**BBKT96**]. Basically a pooling design is transversal if the pools can be partitioned into parts, each of which is a partition of the clone population. Clearly the hypercube design is a special case of transversal designs. Not much has been studied toward this general direction. Relations of this problem to Coding Theory is also specified in [**BBKT96**].

**3.3. Direct Constructions.** Macula [**Mac96, Mac99**] gave the following construction of a $d$-disjunct matrix. Let $\delta(m,d,k)$ be a 01-matrix whose rows are indexed by the $d$-subsets of $[m]$ and whose columns are indexed by the $k$-subsets of $[m]$ where $\frac{m}{2} \geq k > d \geq 1$. $\delta(m,d,k)_{ij} = 1$ iff the $i^{th}$ $d$-subset is contained in the $j^{th}$ $k$-subset. It is easy to see that $\delta(m,d,k)$ is $d$-disjunct with $\binom{m}{d}$ rows and $\binom{m}{k}$ columns.

The number of tests to number of items ratio of $\delta(m,d,k)$ is $\binom{m}{d}/\binom{m}{k}$, which is not so good in terms of the random bound given in Theorem 1. However, Macula showed that with high probability $\delta(m,2,k)$ could solve the $S(\bar{d},n)$ problem, effectively converting a deterministic construction to a probabilistic (random) one. This point will be discussed further in a later section. In addition, $m$ and $k$ could be chosen carefully in certain cases to suit one's need. However, the method of choosing these parameters needs more thorough analysis than just trial and error.

## 4. On Constructions of $d$-disjunct Matrices

In set packing designs, the matrix $M$ was row indexed by all elements of $[v]$, and column indexed by selected $k$-subsets of $[v]$. Looking at this from a different angle, the rows were indexed by all points at rank 1 and columns by sampled points at rank $k$ of the Boolean Algebra lattice $B_v$ (see Figure 1).

On the other hand, Macula's construction involves taking all points at rank $d$ as rows and rank $k$ as columns of our $d$-disjunct matrix. Macula's design rate wasn't so good because number of points at level $d$ is too large. However, if we pick points at lower levels than $d$ to be the rows, then the matrix is not $d$-disjunct anymore.

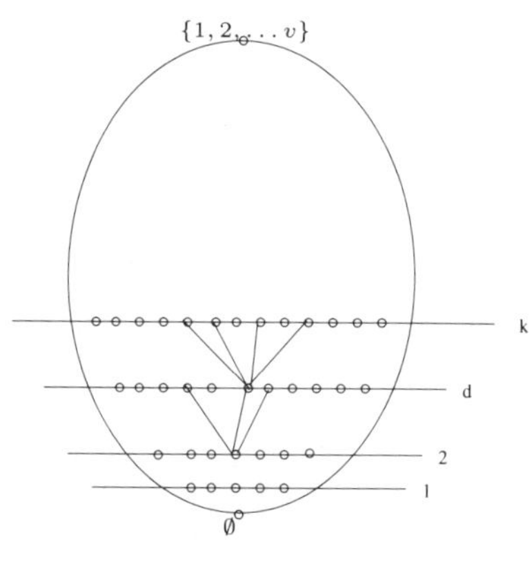


Figure 1: The Boolean Algebra Lattice

Stretching this line of reasoning, one might hope to somehow take sampled points at different ranks of $B_v$, not taking *all* points. Ngo and Du (1999, [**ND99**]) took this approach and gave the following construction. Given integers $m \geq k > d \geq 1$. A matching of size $l$ in $K_v$ is called an $l$-matching. Let $M(m,k,d)$ be a 01-matrix whose rows are indexed by the set of all $d$-matchings on $K_{2m}$, and whose columns are indexed by the set of all $k$-matchings on $K_{2m}$. All matchings are to be ordered lexicographically. $M(m, k, d)$ has a 1 in row $i$ and column $j$ if and only if the $i^{th}$ $d$-matching is contained in the $j^{th}$ $k$-matching. The fact that $M(m, k, d)$ is $d$-disjunct is not difficult to be seen. Noticing that a $k$-matching is a $k$ subset of $\left[\binom{2m}{2}\right]$, because the set of edges of $K_{2m}$ is exactly $\left[\binom{2m}{2}\right]$. From the above observation, this construction could be seen as taking from $B_{[\binom{2m}{2}]}$ sampled points at rank $d$ as rows and sampled points at rank $k$ as columns. Ngo and Du also showed that $M(m, m, d)$ is $d$-error-detecting and $\lfloor \frac{d}{2} \rfloor$-error-correcting.

On another dimension, the Boolean Algebra is clearly not the only lattice we have to work on. Some obvious questions arising would be

(1) Besides the Boolean Algebra $B_v$, what are other lattices we can use ? For example, one candidate is $C_{v,u}$, the lattice of all $v$ tuples of $Z_u$, which is a generalization of $B_v$, since $B_v = C_{v,2}$.
(2) Which conditions must hold to pick some two levels of the lattice to construct $d$-disjunct matrices ? To avoid being too vague and for the ease of analysis, we could restraint ourselves to the lattices with some regularity constraint. An example would be to work on lattices where the number of points covering a point $p$ at rank $k$ and the number of points covered by $p$ depend only on $k$.
(3) In terms of error tolerance properties, can we from the lattice infer some information about the error correcting and detecting capability of the matrix being constructed ?

With respect to question 1, Ngo and Du [**ND99**] found that picking points at levels $d$ and $k$ of the lattice of all subspaces of $\mathbb{F}_q^v$ would also work. This construction, in fact, is the $q$-analog of Macula's construction.

## 5. Random Pooling Designs

Random designs refer to the designs whose matrices are randomly determined in some manner. The fact that a design is nondeterministic means that it is possible for some positives and negatives not to be identified. Let $M$ be a random $v \times n$ matrix, our algorithm of identifying the defectives is the same as before, namely pointing those items contained in negative tests as negative. These are called *resolved negatives*. Clearly, an item in a positive pool where all others in the pool are resolved negatives must be positive. These positive items are said to be *resolved positives*. Let $\bar{N}$ ($\bar{P}$) denote the number of unresolved negatives (positives). Balding et al. introduced several criteria to compare designs such as $P(\bar{N} = \bar{P} = 0)$, $P(\bar{N} = 0)$, $E(\bar{P})$, and $E(\bar{N})$, where $P(X = j)$ is the probability of $X = j$ and $E(X)$ is the expected value of a random variable $X$. We would like the probabilities to be as close to 1 as possible and the expected values to be as small as they can get.

**5.1. Random Matrices.** Erdős (as usual) and Renyi (1963, [**ER63**]) first introduced random methods in search problems. Much later, Sebő (1985, [**Seb85**]) adopted the idea to group testing. To construct a random disjunct matrix $M$, we simply assign 1 to an entry of $M$ with some fixed probability $p$. Given $n$ and $d$,

$p$ and $v$ could be chosen properly so that the probability of $M$ being $d$-disjunct is higher than some certain tolerable threshold.

Although this method is not used in practice, partially due to its bad performance [**BBKT96**], the idea can be used to obtain very good bounds on the number $v(d, n)$. Theorem 1 is an example of such random bounds.

**5.2. Random Weight-$w$ Designs.** If a clone is contained in no pool, we don't have any information above the clone. If a clone is contained in every pool and it happens to be positive, all tests turn out to be positive and thus the amount of information we get is also zero. On the same line of reasoning, a design with a clone contained in too many or too less number of tests is not good. Moreover, if the number of pools containing a clone varies, then the analysis would be very tedious if not impossible. Consequently, it is reasonable to attempt constructing random matrices with some constant weight $w$, where $w$ could be chosen to optimize some of the efficiency criteria. This could be done by assigning the columns randomly to $w$-subsets of $[v]$. These designs are called *random weight-$w$ designs*. Let the corresponding probabilities and expected values be denoted by $P_w(\cdot)$ and $E_w(\cdot)$ respectively. Let $W(i)$ denote the probability that a particular set of $i$ pools is exactly the set of pools not containing any positive clones. The following formulas were obtained by Bruno et al. [**BKB$^+$95**], and Hwang [**Hwa99**].

$$W(i) = \sum_{h=i}^{v} (-1)^{h-i} \binom{v-i}{h-i} \left[ \frac{\binom{v-h}{w}}{\binom{v}{w}} \right]^d \tag{5.1}$$

$$P_w(\bar{N} = j) = \sum_{i=0}^{v} \binom{v}{i} W(i) \binom{n-d}{j} \left[ 1 - \frac{\binom{v-i}{w}}{\binom{v}{w}} \right]^{n-d-j} \left[ \frac{\binom{v-i}{w}}{\binom{v}{w}} \right]^j \tag{5.2}$$

$$E_w(\bar{N}) = (n-d) \sum_{i=1}^{w} \binom{w}{i} \left[ \frac{\binom{v-i}{w}}{\binom{v}{w}} \right]^d \tag{5.3}$$

An open question is to find $w$ so that $E_w(\bar{N})$ is minimized. Notice that these formulas were calculated ignoring the fact that in practice we don't want identical columns in the matrix. The reason is that taking into account this fact makes the calculation more difficult. Bruno et al. [**BKB$^+$95**] also indicated that random weight-$w$ designs perform better than the random design discussed earlier.

**5.3. Random Size-$k$ Designs.** Dually, instead of reasoning on the columns of $M$ we could do the same on the rows of $M$. A pool containing too few clones is wasted if these clones are negatives, while a pool containing too many clones gives little information if there is a positive clone in it. Hence, we could as well randomly choose the rows of $M$ with some constant size $k$ uniformly. Similar formulas as those in the last sections were obtained by Hwang (1999, [**Hwa99**]):

$$(5.4) \qquad P_k(\bar{N} = n - d) = \left[1 - \frac{\binom{n-d}{k}}{\binom{n}{k}}\right]^v$$

$$P_k(\bar{N} = j) = \sum_{i=0}^{v} \binom{v}{i} \left[\frac{\binom{n-d}{k}}{\binom{n}{k}}\right]^i \left[1 - \frac{\binom{n-d}{k}}{\binom{n}{k}}\right]^{v-i} \cdot \sum_{l=j}^{n-d} (-1)^{l-j} \binom{n-d}{l} \left[\frac{\binom{n-d-l}{k}}{\binom{n-d}{k}}\right]^i$$

$$(5.5) \qquad \text{for } 0 \leq j < n - d$$

In the same paper, Hwang also gave formulas to compute $E_x(\bar{P})$ for $x \in \{p, w, k\}$. Here $E_p(X)$ denote the expected value of $X$ when $M$ is constructed using the first random method with probability $p$.

**5.4. Random Designs from Deterministic Designs.** Macula [**Mac99**] showed that his matrix $\delta(m, 2, k)$ could be used to solve the $S(\bar{d}, n)$ problem with high probability of success. Clearly, this is desirable since the test to item ratio of $\delta(m, 2, k)$ is smaller than that of $\delta(m, d, k)$ in general. The probability, denoted by $P_\delta(n, d, k)$, can be shown to be

$$P_\delta(n, d, k) \geq \left[\frac{\sum_{i=1}^{k} (-1)^{i+1} \binom{k}{i} \binom{\binom{n-i}{k}}{d-1}}{\binom{\binom{n}{k}-1}{d-1}}\right]^d$$

For example, when $d = 5$ and $n \approx 1,000,000$ we can pick $\delta(44, 2, 5)$, which has 946 rows (tests), 1,086,008 columns (items), and $P_\delta(44, 5, 5) \geq .97107$.

Borrowing this idea, Ngo and Du [**ND99**] also showed that $M(m, k, 2)$ could be used to solve $S(\bar{d}, n)$ with probability $P_M(n, k, d)$ of giving the right answer, where

$$P_M(m, k, d) \geq \left[\frac{\sum_{j=1}^{k} (-1)^{j+1} \binom{k}{j} \binom{\sum_{i=0}^{j} (-1)^i \binom{j}{i} g(m-i, k-i)}{d-1}}{\binom{g(m,k)-1}{d-1}}\right]^d$$

Here, $g(m, l) = \binom{2m}{2l} \frac{(2l)!}{2^l l!}$. For example, $P_M(8, 6, 9) \geq 98.5\%$, with the number of defectives $d = 9$, the number of items $n = g(8, 6) = 18,918,900$ and the number of test $v = g(8, 2) = 5460$.

One can see from these formulas that the efficiency benchmarks to compare pooling designs often involve complicated, hypergeometric type of formulas arising from inclusion exclusion enumerations. This makes the analysis difficult and tedious. Usually, what we can do is to plug in some particular values and do manual comparison, which is clearly not satisfactory theoretically. More work needs to be done in asymptotic analysis of these formulas in order to give satisfactory results.

## 6. Error Tolerance Pooling Designs

As we have mentioned earlier, when DNA probing could be error prone, which leads us to the greater challenge of designing pools that could tolerate some number of errors. This problem is the non-adaptive version of the *searching game* initiated by Ulam [**Ula76**] back in 1976. Ulam's problem was to determine a chosen number $u$ out of $[n]$ using the minimum number of questions of the form: Is $u \in S$, $S \subseteq [n]$. Moreover, the responder could lie once or twice. In general, the questions and answers could be $q$-ary, i.e. each question is a partition of $[n]$ into $q$ parts and each answer points out which part(s) any of $d$ unknowns belong to. Up to $e$ lies is allowed. It is easy to see that our problem is the non-adaptive version of this so-called *q-ary search problem with lies* where $q = 2$. Although quite a lot of research effort has been put on solving this problem, we only have solutions for several special cases where $q$ and $e$ are small.

Adaptively, when $d = 1, q = 2$ Pelc [**Pel87**] solved the case $e = 1$, Guzicki [**Guz90**] solved the case $e = 2$, and Spencer [**Spe92**] provided a nearly optimal solution (up to a constant) for general $e$. The $q$-ary case (with $d = 1$) was consider by Aigner [**Aig96**] and Muthukrishnan [**Mut94**] with complete solutions.

Non-adaptively, several author have noticed that when $d = 1$, the design is equivalent to an $e$-error-correcting code. Balding and Torney [**BT96**] studied several instances of the problem when $d \leq 2$. They showed that an optimal strategy is possible if and only if certain Steiner system exists. Macula [**Mac99**] showed that his construction is error tolerable up to certain calculatable probability. Ngo and Du construction [**ND99**] was shown to be $d$-error-detecting and $\lfloor \frac{d}{2} \rfloor$-error-correcting in the worst case, but can tolerate more errors on average.

We need deeper results and new breakthroughs in order to improve our present knowledge of the most general case of the problem, especially in the non-adaptive scenario. For example, we need good bounds similar to those in Theorem 1 given the number of items $n$, maximum number of defectives $d$ and maximum number of errors $e$.

## 7. Conclusions

In this paper, we have given an overview of up-to-date results on Combinatorial Group Testing algorithms which are applicable to DNA library screening. We have been focusing more on new classes of constructions not previously discussed and pointed out directions to generalize existing results. We also have discussed some related open questions popped up in this area.

Finally, we would like to conclude that this is a young and interesting field with deep connections to Coding Theory and Design Theory. We strongly believe that the theory Distance Regular Graphs, in particular Association Schemes, should play an important role in improving our pooling designs.

## References

[Aig96] Martin Aigner, *Searching with lies*, J. Combin. Theory Ser. A **74** (1996), no. 1, 43–56.

[BBKT96] D. J. Balding, W. J. Bruno, E. Knill, and D. C. Torney, *A comparative survey of non-adaptive pooling designs*, Genetic mapping and DNA sequencing (Minneapolis, MN, 1994) (New York), Springer, New York, 1996, pp. 133–154.

[BJL86] Thomas Beth, Dieter Jungnickel, and Hanfried Lenz, *Design theory*, Cambridge University Press, Cambridge, 1986.

[BKB+95] W. J. Bruno, E. Knill, D. J. Balding, D. C. Bruce, N. A. Doggett, W. W. Sawhill, R. L. Stallings, C. C. Whittaker, and D. C. Torney, *Efficient pooling designs for library screening*, Genomics (1995), no. 26, 21–30.

[BLC91] E. Barillot, B. Lacroix, and D. Cohen, *Theoretical analysis of library screening using a $n$-dimensional pooling strategy*, Nucl. Acids Res. (1991), no. 19, 6241–6247.

[Bro79] A. E. Brouwer, *Optimal packings of $K_4$'s into a $K_n$*, J. Combin. Theory Ser. A **26** (1979), no. 3, 278–297.

[BT96] David J. Balding and David C. Torney, *Optimal pooling designs with error detection*, J. Combin. Theory Ser. A **74** (1996), no. 1, 131–140.

[DH93] Ding Zhu Du and Frank K. Hwang, *Combinatorial group testing and its applications*, World Scientific Publishing Co. Inc., River Edge, NJ, 1993.

[DR82] A. G. Dyachkov and V. V. Rykov, *Bounds on the length of disjunctive codes*, Problemy Peredachi Informatsii **18** (1982), no. 3, 7–13.

[DR83] A. G. Dyachkov and V. V. Rykov, *A survey of superimposed code theory*, Problems Control Inform. Theory/Problemy Upravlen. Teor. Inform. **12** (1983), no. 4, 229–242.

[DRR89] A. G. Dyachkov, V. V. Rykov, and A. M. Rashad, *Superimposed distance codes*, Problems Control Inform. Theory/Problemy Upravlen. Teor. Inform. **18** (1989), no. 4, 237–250.

[ER63] Paul Erdős and Alfréd Rényi, *On two problems of information theory*, Magyar Tud. Akad. Mat. Kutató Int. Közl. **8** (1963), 229–243.

[Guz90] Wojciech Guzicki, *Ulam's searching game with two lies*, J. Combin. Theory Ser. A **54** (1990), no. 1, 1–19.

[Hwa95] F. K. Hwang, *An isomorphic factorization of the complete graph*, J. Graph Theory **19** (1995), no. 3, 333–337.

[Hwa99] F. K. Hwang, *Random size-$k$ pool designs with distinct columns*, preprint.

[KS64] W. H. Kautz and R. C. Singleton, *Nonrandom binary superimposed codes*, IEEE Trans. Inf. Theory **10** (1964), 363–377.

[Mac96] Anthony J. Macula, *A simple construction of $d$-disjunct matrices with certain constant weights*, Discrete Math. **162** (1996), no. 1-3, 311–312.

[Mac99] Anthony J. Macula, *Probabilistic nonadaptive group testing in the presence of errors and dna library screening*, Annals of Combinatorics (1999), no. 3, 61–69.

[MM92] W. H. Mills and R. C. Mullin, *Coverings and packings*, Contemporary design theory (New York), Wiley, New York, 1992, pp. 371–399.

[Mut94] S. Muthukrishnan, *On optimal strategies for searching in presence of errors*, Proceedings of the Fifth Annual ACM-SIAM Symposium on Discrete Algorithms (Arlington, VA, 1994) (New York), ACM, 1994, pp. 680–689.

[ND99] H. Q. Ngo and D. Z. Du, *New constructions of non-adaptive and error-tolerance pooling designs*, preprint.

[OHCB89] M. Olson, L. Hood, C. Contor, and D. Botstein, *A common language for physical mapping of the human genome*, Science (1989), no. 245, 1434–1435.

[Pel87] Andrzej Pelc, *Solution of Ulam's problem on searching with a lie*, J. Combin. Theory Ser. A **44** (1987), no. 1, 129–140.

[Rom92] Steven Roman, *Coding and information theory*, Springer-Verlag, New York, 1992.

[Sch66] J. Schönheim, *On maximal systems of $k$-tuples*, Studia Sci. Math. Hungar **1** (1966), 363–368.

[Seb85] András Sebő, *On two random search problems*, J. Statist. Plann. Inference **11** (1985), no. 1, 23–31.

[Spe92] Joel Spencer, *Ulam's searching game with a fixed number of lies*, Theoret. Comput. Sci. **95** (1992), no. 2, 307–321.

[Ula76] S. M. Ulam, *Adventures of a mathematician*, Charles Scribner's Sons, New York, 1976.

DEPARTMENT OF COMPUTER SCIENCE AND ENGINEERING, UNIVERSITY OF MINNESOTA,, 200 SE UNION ST., 4-192 EE/CS BLDG, MINNEAPOLIS, MN 55455
*E-mail address*: hngo@cs.umn.edu

DEPARTMENT OF COMPUTER SCIENCE AND ENGINEERING, UNIVERSITY OF MINNESOTA,, 200 SE UNION ST., 4-192 EE/CS BLDG, MINNEAPOLIS, MN 55455
*E-mail address*: dzd@cs.umn.edu

DIMACS Series in Discrete Mathematics
and Theoretical Computer Science
Volume **55**, 2000

# Optimization Problems in 3D Conformal Radiation Therapy

Q. Jackie Wu, Jie Wang, and Claudio H. Sibata

Abstract. Three-dimensional (3D) conformal radiation therapy is a complex optimization process with the following goal: Maximize dose distribution to the target volume and, at the same time, minimize toxicity to the surrounding normal tissues and critical organs. To achieve this goal we need to select proper beam parameters, find objective functions that can incorporate and formulate physician's judgment of a treatment plan, and devise efficient algorithms to find a solution of the objective functions. Many common objective functions are discrete combinatorial optimization problems. Thus, finding efficient algorithms to solve such problems would be an interesting topic in computer science. This paper is intended to provide a survey of the optimization problems in 3D conformal radiation therapy for mathematicians and computer scientists interested in this area.

## 1. Introduction

In conformal radiation therapy, we want to deliver the spatial distribution of the prescribed dose to the target volume such that the dose delivered to the surrounding normal tissues and critical organs is minimized. Conformal radiation therapy has two advantages over conventional radiation therapy. First, it can effectively exclude radiation sensitive organs to allow us to increase tumor dose to levels that conventional radiation therapy cannot reach, while decreasing the risk of radiation induced morbidity. Second, it can take advantage of dose escalation to significantly increase local control rates for a variety of tumors.

Thanks to the advanced technologies of computer hardware and software, sophisticated 3D radiation therapy treatment-planning systems and treatment-delivery systems have been developed based on modern linear accelerators (Linac). Linac provides 3D translational and rotational movements around the patient, so that the direction of a beam can have an arbitrary angle in 3D space. These movements can be made automatic either in a single treatment field (beam), or among treatment fields. Fig. 1 shows the diagram of a Linac, where the gantry, the collimator, and the patient couch can be positioned independently, providing six-degree freedom of beam orientation. Moreover, the Multi-Leaf Collimator (MLC), which is housed in the gantry head, can provide high spatial and intensity resolution within each beam. The MLC is a set of narrow tungsten leaves (typically projecting 1 cm wide

2000 *Mathematics Subject Classification.* Primary 92C50, 68W99, 68R05; Secondary 90C90.
The second author was supported in part by NSF under grant CCR-9820611.

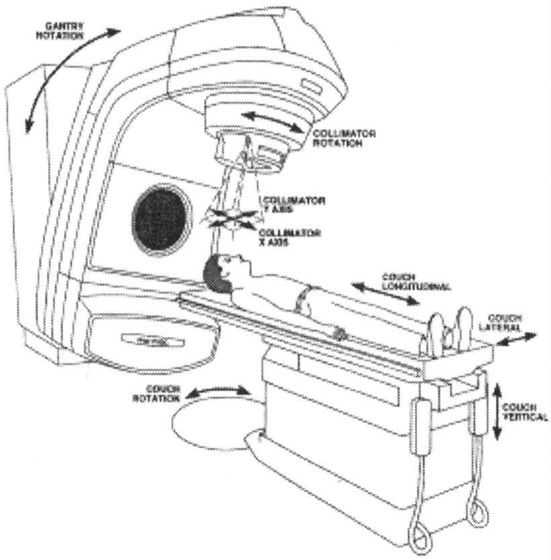


FIGURE 1. Diagram of a Linac.

FIGURE 2. Multi-Leaf Collimator (MLC): Leaves are positioned to match geometrically the shape of the projected target area.

at the “isocenter”) that can be individually positioned and moved under computer control during the irradiation of the beam. Fig. 2 shows the MLC that is positioned to shape a treatment field geometrically so that only the projected target area will be irradiated. This shaped beam or field can be further divided into a matrix, or beamlets, as shown in Fig. 3. The radiation intensity of each beamlet can be further manipulated by controlling the opening time of the individual MLC leaf. Fig. 3 also illustrates a radiation intensity map delivered by the MLC.

3D conformal treatment planning and conventional radiation therapy planning differ in two major aspects. First, the planning of 3D conformal therapy is volumetric based. For conformal treatment and planning, patient anatomical data are obtained using CT or MRI. Beam geometries and treatment fields are designed based on the irradiation of a 3D-target volume in the 3D space, using Beam’s-Eye-View and 3D graphical displays. Dose distribution is calculated in 3D, and the treatment plan is analyzed and evaluated in 3D using dose-volume-histograms (DVH) or other score functions calculated in the 3D volume of interest. In conventional

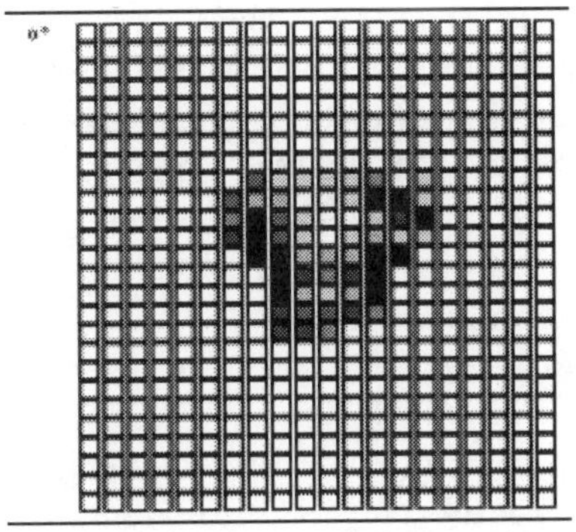

FIGURE 3. Multi-Leaf Collimator (MLC): A radiation intensity map generated by manipulating individual leaf's opening time.

treatment plan evaluation, however, the dose distribution is simply superimposed on transverse planes. Second, 3D conformal radiation therapy has renewed interests in automatic computer optimization to improve efficiency and to deal with a very large base of beam parameters. Automatic computer optimization involves the following two components:

1. an objective function or score function that provides a figure of merit of a particular plan; and
2. an optimization algorithm that computes the optimal plan.

Such an optimization process is often refereed to as *inverse planning*; while in conventional radiation therapy, *forward planning* is used, where treatment field parameters are defined before the planning starts.

Ideally, the parameters that need to be optimized should include the total number of beams used, the orientation, energy, shape of each beam, and the intensity of the beamlets within each beam. Finding efficient algorithms and developing computer systems to solve this optimization problem remain a challenge. Currently, most 3D conformal radiation therapies are accomplished mainly by shaping the incident beam fields in spatial and/or intensity distribution for a set of fixed beam direction.

3D conformal radiation therapy optimization is an interactive process in which the planner plays a vital role in designing the optimization goals. If unrealistic constraints are given, the chance to obtain a clinically satisfactory plan is likely to be small.

This paper is structured as follows. A brief introduction of conventional radiation therapy is given in Section 2. Some of the commonly used optimization techniques are discussed in Section 3. We present in Section 4 some of the objective goals one would like to implement in clinical conformal treatment, and discuss in Section 5 some future research topics in the field.

## 2. Conventional Radiation Treatment and Planning

Simulation is the key component of conventional radiation treatment and planning. It uses a simulator and a gantry mounted x-ray tube/imaging system (film or film plus imaging intensified fluoroscopy) to perform treatment and planning simulations. The simulator has the same movements as a Linac to reproduce the field setups for the treatment. The x-ray source is a conventional x-ray tube. The x-ray films taken at the simulator are used to compare with portal films taken at a Linac for verification of the clinical setup and treatment.

The simulation process involves the following steps.

1. Immobilize the patient. The patient has to be placed on the simulator table in the exact same position on a Linac. Various immobilization devices are used to ensure daily reproducibility.
2. Establish a coordinate system. Once the beam placement is configured, a coordinate system is established to treat the patient in a consistent manner. Localization lasers are used to mark the patient, where the patient is properly localized with the isocenter corresponding to the simulator isocenter. These setup marks are then used to align the patient target to the isocenter on the Linac he/she is using.
3. Import patient contour and target information. In conventional simulation the external contour of the patient are obtained manually using wires or contour tracing device. With the advantage of the computed tomography (CT), this anatomical information can be directly imported into a treatment planning system.
4. Design treatment plans. In conventional radiation therapy, the simulation process is used to obtain beam configuration; so the treatment planning is limited to calculating the dose distribution based on the patient anatomical information relative to the beam configuration. The beam shapers or modifiers can be placed in the beam to improve the dose homogeneity in the target.

Conventional treatment planning is based on the assumption that we know how to treat the disease; i.e., we know *a priori* what the beam arrangements will be. The simulation process is simply to find the treatment isocenter and the proper beam size to cover the target based on the bony landmarks. This is an experience-based, trial-and-error approach: The planning software does not actually plan; rather, it simply simulates a user-defined set of beams and beam weights to search for a suitable plan.

As treatment planning systems are developed to include CT images and image processing tools, it becomes possible to simulate both the patient and the treatment machine in virtual space. Picker International was the first to develop such a system for commercial use. The system, called AcQsim, uses image processing techniques to get Beam's-Eye-View images, determine beam shapes based on the projection of the target volume, and generate Digitally-Reconstructed-Radiograph (DRR) to replace the simulation films. The process, however, expects to obtain beam arrangements from the user, and so it is a forward-planning process.

Inverse planning is a target-driven treatment planning process, which takes the patient anatomical information as input and outputs a treatment field configuration that satisfies the physician's prescription or is close to the prescription as much as possible.

## 3. Optimization Methods

Defining effective cost functions is a complex task, which depends on many issues. Among all possible cost functions, the most common and simplest cost function would be to minimize the errors between the dose desired at each voxel in the treatment volume and the dose that would be actually delivered by the treatment plan.

Let $D_0(n)$ denote the desired dose at voxel $n$ in the 3D dose matrix. The goal here is to find a set of beam parameters that can "most closely" reproduce $D_0(n)$. Mathematically, we can formulate the optimization as a problem of finding $D_c(n)$ such that the following objective function is minimized:

$$F = \sum_{n=1}^{N} r_\sigma [D_c(n) - D_0(n)]^2, \tag{3.1}$$

where $r_\sigma$ is the relative importance factor for structure $\sigma$ to receive the desired dose. We will use this cost function to illustrate some of the optimization methods. The forms of these objective functions also affect the choice and applicability of the optimization methods.

In principle, we can vary the total number of beams, the orientation, the energy, the shape, and the beamlet weight of each beam to generate a different plan [**17**]. Since there are infinitely many combinations of beam parameters, it makes sense to focus on optimizing individual beam parameters, especially the number of beams and the beam directions [**6, 11, 30**]. Some methods use geometric strategies, in which the fractional volumes of normal and critical structures intercepted by radiation beams encompassing the chosen target are plotted as a function of the gantry angle, and are used to select beam directions to reduce the volumes of normal tissues irradiated. In practice, the beam weight and beamlet weights are commonly used as optimization parameters, while all other beam parameters are pre-selected based on intuitions, experiences, and 3D visualization tools.

Let $n$ be a point in the does matrix. Let $J$ be the total number of beams used in the plan. Each beam at a given orientation $\Phi_i$ has $M$ pencil beams of finite size, called beamlets. The size of the beamlet often corresponds to the resolution of the MLC leaf width. Let $D_{jm}$ denote the dose distribution of the $m$th beamlet of the $j$th beam, and $w_{jm}(n)$ the weight of the $m$th beamlet of the $j$th beam. Then $D_c(n)$ is defined as follows:

$$D_c(n) = \sum_{j=1}^{J} \sum_{m=1}^{M} w_{jm}(n) D_{jm}. \tag{3.2}$$

The pencil beam dose distribution $D_{jm}$ can be pre-calculated by convolution or scatter integration algorithms. The planning or optimization process is to search for $w_{jm}(n)$ that minimizes the cost function. The weight can then be translated into relative opening times and movements of each individual MLC leaf for a given beam to deliver the intensity modulated treatment.

Several mathematical and physical techniques have been used to find acceptable solutions. Among them are analytical techniques, filtered-back-projection techniques, and iterative techniques.

The analytical techniques, such as linear and quadratic programming, have been used to find $D_c(n)$ for minimizing $F$ with certain success [**7, 20, 27, 29**]. However, it is not always possible to achieve a perfect inverse solution by analytic means unless negative beam weights are used. Negative beam weights are physically unrealizable.

The filtered-back-projection technique [**1, 12**] is applicable only for optimizing beamlet intensities based on the notion of CT reconstruction. In computed tomography, the radiation transmission profiles at each gantry angle are back-projected to obtain the density distribution of the body. When applied to radiation therapy, this technique takes a user specified dose distribution and inverts it to obtain beamlet intensity profiles at each beam angle placed around the target volume. The resultant distribution, called the *fluence intensity distribution*, leads to a dose distribution in the forward dose calculation that is as close as what can be physically achieved to the desired dose distribution.

Although the back projection for CT reconstruction corresponds to real physical situations of which the images are to be reconstructed, in radiation therapy, the desired dose distribution may or may not be physically realizable. Hence, the derived solution may contain negative beam weights [**18**]. Another disadvantage of the filtered-back-projection technique is that it is only applicable to co-planner beams (i.e., no couch angles are allowed) and the objectives of the optimization are limited to dose distributions.

The iterative technique has been studied intensively (e.g., see [**5, 16, 17, 19, 28, 31, 34, 35, 37**]). The basic idea here is to iteratively correct the beam weights $w_{jm}$ in an attempt to match the calculated dose $D_c(n)$ to the desired dose $D_0(n)$. The iterative process first assigns the initial pencil beam weights $w_{jm}^0$. It then uses a forward-dose-calculation algorithm to compute dose at each voxel $n$. It compares the calculated and desired doses and corrects the weight. In particular, each pencil beam weight is updated as follows:

$$w_{jm}^{new} = w_{jm} + \delta \Delta w_{jm}, \tag{3.3}$$

where $\delta$ is a relaxation parameter used by different optimization schemes. Keep iterating this process until an acceptable solution is obtained, i.e., until the value of the obtained cost function is smaller than a pre-determined tolerance value.

Several methods, such as the direct iteration and the simulated annealing methods, have been used to update $w_{jm}$. The method of direct iteration updates the weights using

$$w_{jm}^{new} = w_{jm} + \Delta w_{jm}. \tag{3.4}$$

This method can be fast, but it can be trapped in a local minimum as well. This can happen when the objective function has too many constraints [**8**]. One way to avoid such problems is to use the simulated annealing method, which mimics the way a thermal system reaches its ground state as the temperature slowly decreases. At each iteration, a small change, either positive or negative, is attempted for one of the beamlet weights. The simulated annealing method makes such an update possible to allow the change of the cost function towards a larger value, which prevents a solution from being trapped in local minima. But this method requires a very large number of iterations [**37**].

Other methods using genetic algorithms and artificial intelligence techniques have also being investigated [**9, 14, 23**]. These algorithms are able to optimize the treatment plan with physical dose, dose-volume limits, and biological end-points such as tumor-control-probability (TCP) and normal-tissue-complication-probability (NTCP). These methods allow the specification to be a combination of dose-volume limits intuitive to the planner, and desirable TCP and NTCP values gauging the ultimate outcomes of the treatment. The specification of dose-volume constraints also provides greater flexibility of planning, for some hot spots may be allowed if the organ-at-risk value is below a specified limit. However, the efficiency of these methods need to be greatly improved.

To end this section, we define the following optimization problem based on the cost function defined in (3.1) and (3.2). This is an easier optimization problem whose solution may be useful to help find a good algorithm to minimize $F$ defined in Equation (3.1).

BEST VALUE MATCHING (BVM)

*Input*: A class of sets $W = \{w_1, \dots, w_n\}$ and a positive integer $D$, where $w_i$ is a finite set of positive integers, $1 \leq i \leq n$.

*Output*: A set of integers $\{v_1, \dots, v_n\}$, where $v_i \in w_i$, such that $(\sum_{i=1}^{n} v_i - D)^2$ is minimized.

## 4. Optimization Goals

Evaluation of a treatment plan by a physician is often a subjective and qualitative process based on standards of the practice and the physician's experience. These factors may include the extent of the disease, perceived probability of tumor control and normal tissue complications, and other clinical factors such as patient status prior to radiation, etc. For computerized treatment planning optimization, these factors must be defined quantitatively, in the form of an objective function or a score function that can be mathematically formulated and evaluated. Ideally, such a function should incorporate mathematical representations of all aspects, clinical, biological, and others, of a patient's information. Unfortunately, not much is known on how to define such functions. Simplified forms of objective functions are used in current practice, which are only a subset of these factors [**17**].

Various forms of objective functions and constraints have been proposed and investigated (e.g., see [**2, 3, 7, 8, 10, 14, 15, 22, 32, 33, 36**]). If the form of the objective function and the constraints are limited to be linear or quadratic, then linear or quadratic programming methods may be applicable. Examples of such limits include maximizing dose to the tumor; minimizing dose to the critical organs and normal structures; and minimizing dose variation over the target volume, or the sum of least square deviation of desired dose distribution at points within the target or critical structures. In addition, the user can have more control over the "balance" between the objectives by assigning importance factors $r_\alpha$ to different regions or structures [**38**]. For example, the dose points can be grouped into plan-target-volume (PTV) points, and a series of organ-at-risk value (OAR) points. A generic least-square-type objective function can be written as:

$$F = S_1 + S_2, \tag{4.1}$$

$$S_1 = \sum_{n \in PTV}^{N} r_{PTV}[D_c(n) - D_0(n)]^2, \tag{4.2}$$

$$S_2 = \sum_{i=1}^{K} \left( \sum_{n \in OAR_i}^{N} r_{OAR_i}[D_c(n) - D_0(n)]^2 \right)^2, \tag{4.3}$$

where $r_{PTV}$ is the importance factor for the PTV and there are $K$ OAR sets with importance factors $r_{OAR_i}, i = 1, \dots, K$, respectively. $D_0(n)$ is the prescribed dose distribution over the 3D matrix of the dose points and $D_c(n)$ is the calculated dose distribution of the current computer generated plan.

We know, however, that the response of the tumor and normal tissues is a function of not just dose but also of the volume receiving the dose at each level [**20, 21**]. Thus, dose volume constraints are often used as prescription and constraints. Dose volume considerations are typically stated as "no more than $x\%$ of volume of organ $i$ may receive a dose greater than dose $D_y$", or "no less than $m\%$ of volume of the target may receive a dose less than dose $D_n$". Many layers of the dose volume constraints may be prescribed to a specific structure.

Using models supported by observational data, biological responses such as TCP and NTCP can be computed from the 3D dose distributions in the PTV and OARs as follows.

$$TCP = \sum_{n=1}^{N} f_{TCP,n\in PTV}(D(n)) \tag{4.4}$$

$$NTCP_{OAR_i} = \sum_{n=1}^{N} f_{NTCP,n\in OAR_i}(D(n)) \tag{4.5}$$

where $f_{TCP,n\in PTV}$ represents the functional model which operates on the dose points $n$ who are members of PTV. The dose distribution can be inhomogeneous. Also, $f_{NTCP}$ represents the functional model, which operates on the dose point n that is members of the $OAR_i$.

Clinical and laboratory data suggest that both target and normal tissues have a sigmoid dose-response relationship in several functional forms [**4**]. For example, the normal probability or profit function model is given as

$$P(D,1) = 0.5\left(1 - erf\left[\sqrt{\pi}\gamma_n\left(1 - \frac{D}{D_{50}}\right)\right]\right); \tag{4.6}$$

while the logistic function model is given as

$$P(D,1) = \left[1 + \left(\frac{D_{50}}{D}\right)^{4\gamma_n}\right]^{-1}; \tag{4.7}$$

and the Poisson statistics function for cell killing model is given as

$$P(D,1) = 2^{-e^{[e\gamma_n(1-D/D_{50})]}}. \tag{4.8}$$

In all these response functions, $D_{50}$ is the dose that yields a 50% response, and $\gamma_n$ is a dimensionless number equal to the maximum value of the normalized gradient $\gamma = D \times dP/dD$. Although these response functions have different forms, they can be made to correspond to a suitable choice of parameters.

It is more complicated to calculate the tissue response probability due to the following situations: Organs or structures are often irradiated with a non-uniform dose distribution and the dose-volume response of the tissues needs to be properly modeled [**13, 24, 25, 26, 36, 39**]. Moreover, these forms depend on basic assumptions about the architecture of the organ in terms of the arrangement of the functional subunits. The behavior of tissue damage from radiation can often be characterized by two simplified models: the serial model and the parallel model. In the serial model, the structure is idealized as a "chain" of functional subunits. If any of the functional subunits is inactivated, the function of the structure is destroyed as a whole. Spinal cords, nerves, and bowels are examples of organs that fall into this category. In the parallel mode, it is assumed that the structure is functional if a significant percentage of the functional subunits remains intact; i.e., the structure can sustain some serious damage to a large percentage of its volume and continue to function normally. Kidneys, livers, and lungs are examples of parallel organs.

It is usually assumed that all functional subunits must be inactivated in order to destroy a tumor. Thus, from a structural point of view, the tumor is a tissue with a parallel architecture. If we divide up the tumor volume into a number of tumorlets or small volumes in which the dose can be assumed to be constant, the overall TCP for a dose-volume distribution curve is simply the product of the response probability of the tumorlets:

$$TCP = \prod_i P(D_i, V_i), \tag{4.9}$$

where $P(D_i, V_i)$ denotes the response probability of the $i$th fractional volume $V_i$ irradiated to the dose $D_i$.

The NTCP for structures that exhibit serial functional architecture can be given as follows:

$$NTCP = 1 - \prod_i (1 - P_i(D, 1))^v. \tag{4.10}$$

These are simplistic models for predicting TCP and NTCP, for the biological structures are more complex than simple serial or parallel models. Also, they were derived from poorly documented and sparse data, and have many deficiencies. Hence, their computed values cannot be taken seriously. However, if they are viewed as relative indices for condensing dose-volume effects to represent the equivalent clinical effect, they are useful for comparing rival plans and for extrapolating conventional experience to the unusual dose distributions observed and anticipated in 3D conformal therapy.

## 5. Final Remarks

In general, the optimization of a 3D conformal treatment plan includes the following steps:

1. Select the optimal number of beams and locations of these beams.
2. Design the shape for each beam.
3. Determine the relative beam weight.
4. Determine the relative beamlet intensity within a beam.

This optimization process could create a huge hyperspace of beam parameters. To reduce the search space, one would like to keep the problem relatively less sensitive to the choice of beam paramagnets. This is one of the motivations for including beamlet intensity modulation into the optimization process. By placing a large number of beams all around the patient and allowing the intensity of each beamlet to be optimized, some researchers argued that there is no need to find the best choice of other beam parameters. However, others find that there are many clinical cases in which a careful selection of incident beam directions can make a signification difference in the final dose distribution, even when only nine incident beams are used. How to select the most important beam parameters and to condense the optimization search space remains to be explored. Optimization has played an important role in advancing the 3D conformal therapy and in making it more efficient. Many optimization methods have been proposed for a wide range of optimization goals, but there are no clear winners yet. When the optimization is expressed in a more clinically relevant manner, the search space may have multiple minima. Therefore, mathematical methods of optimization that allow escapes for local minima should have an advantage over other techniques.

Most objective functions that have been used clinically are based on dose-volume criteria such as minimum dose to the target volume or does-volume constraint to the critical organs and normal structures. Hybrid objective functions seem to be very useful compromises between the natural desire to specify biological endpoints (but with potentially uncertain data and/or model) and the traditional (though surrogate) use of dose distribution, especially when the 3D conformal therapy is still a learning curve for many physicians and researchers. Objective functions will eventually be expressed in terms of biologic indices such as TCP and NTCP. The competition between tumor control and normal organ toxicity still needs to be addressed. The development of robust TCP and NTCP models that accurately predict clinical outcome will emerge as dose-volume outcome data are gathered and analyzed.

The current process of deciding between two competing plans is also an area where tools can be improved. Making a good decision is often difficult. Sometimes it is because we simply do not know which of the followings is better for a particular structure: more tissue exposed to a low dose or a lesser amount of tissue exposed to a higher dose. Sometimes it is because there is too much information to process (multiple sensitive structures each receiving a different dose as determined by multiple plans).

## References

[1] T. Bortfeld, J. Burkelbach, R. Boesecke, and W. Schlegel. Methods of image reconstruction from projections applied to conformation radiotherapy. *Phys Med Biol*, 35(10):1423–1434, 1990.

[2] T. Bortfeld. Optimized planning using physical objectives and constraints. *Semin Radiat Oncol*, 9(1):20–34, 1999.

[3] A. Brahme. Optimized radiation therapy based on radiobiological objectives. *Semin Radiat Oncol*, 9(1):35–47, 1999.

[4] C. Burman, G.J. Kutcher, B. Emami, and M. Goitein. Fitting of normal tissue tolerance data to an analytic function. *Int J Radiat Oncol Biol Phys*, 21(1):123–135, 1991.

[5] R.E. Cooper. A gradient method of optimizing external-beam radiotherapy treatment plans. *Radiology*, 128(1):235–243, 1978.

[6] S.K. Das and L.B. Marks. Selection of coplanar or noncoplanar beams using three-dimensional optimization based on maximum beam separation and minimized nontarget irradiation. *Int J Radiat Oncol Biol Phys*, 1; 38(3):643–655, 1997.

[7] C. DeWagter, C.O. Colle, L.G. Fortan, B.B. Van Duyse, D.L. Van den Berge, and W.J. DeNeve. 3D conformal intensity-modulated radiotherapy planning: interactive optimization by constrained matrix inversion. *Radiother Oncol*, 47(1):69–76, 1998.

[8] J.O. Deasy. Multiple local minima in radiotherapy optimization problems with dose-volume constraints. *Med Phys*, 24(7):1157–1161, 1997.

[9] G.A. Ezzell. Genetic and geometric optimization of three-dimensional radiation therapy treatment planning. *Med Phys*, 23(3):293–305, 1996.

[10] M.V. Graham, N.L. Jain, M.G. Kahn, R. Drzymala, and J.A. Purdy. Objective plan evaluation using a score function tool. *Front Radiat Ther Oncol*, 29:81–85, 1996.

[11] O.C. Haas, K.J. Burnham, and J.A. Mills. Optimization of beam orientation in radiotherapy using planar geometry. *Phys Med Biol*, 43(8):2179–2193, 1998.

[12] T. Holmes and T.R. Mackie. A comparison of three inverse treatment planning algorithms. *Phys Med Biol*, 39(1):91–106, 1994.

[13] A. Jackson, G.J. Kutcher, and E.D. Yorke. Probability of radiation-induced complications for normal tissues with parallel architecture subject to non-uniform irradiation. *Med Phys*, 20(3):613–625, 1993.

[14] N.L. Jain and M.G. Kahn. Ranking radiotherapy treatment plans using decision-analytic and heuristic techniques. *Comput Biomed Res*, 25(4):374–383, 1992.

[15] G.J. Kutcher. Quantitative plan evaluation: TCP/NTCP models. *Front Radiat Ther Oncol*, 29:67–80, 1996.

[16] G.S. Mageras and R. Mohan. Application of fast simulated annealing to optimization of conformal radiation treatments. *Med Phys*, 20(3):639–647, 1993.

[17] R. Mohan, G.S. Mageras, B. Baldwin, L.J. Brewster, G.J. Kutcher, S. Leibel, C.M. Burman, C.C. Ling, and Z. Fuks. Clinically relevant optimization of 3D conformal treatments. *Med Phys*, 19(4):933–944, 1992.

[18] R. Mohan, X. Wang, A. Jackson, T. Bortfeld, A.L. Boyer, G.J. Kutcher, S.A. Leibel, Z. Fuks, and C.C. Ling. The potential and limitations of the inverse radiotherapy technique. *Radiother Oncol* 1994 Sep;32(3):232–48

[19] S.M. Morrill, K.S. Lam, R.G. Lane, M. Langer, and Rosen II. Very fast simulated reannealing in radiation therapy treatment plan optimization. *Int J Radiat Oncol Biol Phys*, 31(1):179–88, 1995.

[20] S.M. Morill, R.G. Lane, J.A. Wong, and Rosen II, Dose-volume considerations with linear programming optimization. *Med Phys*, 18(6):1201–1210, 1991.

[21] S.M. Morrill, Rosen II, R.G. Lane, and J.A. Belli. The influence of dose constraint point placement on optimized radiation therapy treatment planning. *Int J Radiat Oncol Biol Phys*, 19(1):129–141, 1990.

[22] J.E. Munzenrider, A.P. Brown, J.C. Chu, L.R. Coia, K.P. Doppke, B. Emami, G.J. Kutcher, R. Mohan, J.A. Purdy, and B. Shank. Numerical scoring of treatment plans. *Int J Radiat Oncol Biol Phys*, 21(1):147–163, 1991.

[23] A. Niemierko. Random search algorithm (RONSC) for optimization of radiation therapy with both physical and biological end points and constraints. *Int J Radiat Oncol Biol Phys*, 23(1):89–98, 1992.

[24] A. Niemierko. Radiobiological models of tissue response to radiation in treatment planning systems. *Tumori*, 84(2):140–143, 1998.

[25] A. Niemierko and M. Goitein. Calculation of normal tissue complication probability and dose-volume histogram reduction schemes for tissues with a critical element architecture. *Radiother Oncol*, 20(3):166–176, 1991.

[26] A. Niemierko and M. Goitein. Implementation of a model for estimating tumor control probability for an inhomogeneously irradiated tumor. *Radiother Oncol*, 29(2):140–147, 1993.

[27] A.T. Redpath, B.L. Vickery, and D.H. Wright. A new technique for radiotherapy planning using quadratic programming. *Phys Med Biol*, 21(5):781–791, 1976.

[28] Rosen II, K.S. Lam, R.G. Lane, M. Langer, and S.M. Morrill. Comparison of simulated annealing algorithms for conformal therapy treatment planning. *Int J Radiat Oncol Biol Phys*, 33(5):1091–1099, 1995.

[29] Rosen II, R.G. Lane, S.M. Morrill, and J.A. Belli. Treatment plan optimization using linear programming. *Med Phys*, 18(2):141–152, 1991.

[30] S.L. Sailer, J.G. Rosenman, J.R. Symon, T.J. Cullip, and E.L. Chaney. The tetrad and hexad: maximum beam separation as a starting point for noncoplanar 3D treatment planning: prostate cancer as a test case. *Int J Radiat Oncol Biol Phys*, 30(2):439–446, 1994.

[31] O.A. Sauer, D.M. Shepard, and T.R. Mackie. Application of constrained optimization to radiotherapy planning. *Med Phys*, 26(11):2359–2366, 1999.

[32] S. Shalev, L. Bartel, P. Therrien, P. Hahn, and M. Carey. The objective evaluation of alternative treatment plans: I. Images of regret. *Int J Radiat Oncol Biol Phys*, 15(3):763–767, 1988.

[33] S. Soderstrom, A. Gustafsson, and A. Brahme. The clinical value of different treatment objectives and degrees of freedom in radiation therapy optimization. *Radiother Oncol*, 29(2):148–163, 1993.

[34] S.V. Spirou and C.S. Chui. A gradient inverse planning algorithm with dose- volume constraints. *Med Phys*, 25(3):321–333, 1998.

[35] G. Starkschall. A constrained least-squares optimization method for external beam radiation therapy treatment planning. *Med Phys*, 11(5):659–665, 1984.

[36] X.H. Wang, R. Mohan, A. Jackson, S.A. Leibel, Z. Fuks, and C.C. Ling. Optimization of intensity-modulated 3D conformal treatment plans based on biological indices. *Radiother Oncol*, 37(2):140–152, 1995.

[37] S. Webb. Optimization by simulated annealing of three-dimensional, conformal treatment planning for radiation fields defined by a multileaf collimator: II. Inclusion of two-dimensional modulation of the x-ray intensity. *Phys Med Biol*, 37(8):1689–1704, 1992.

[38] L. Xing, J.G. Li, S. Donaldson, Q.T. Le, and A.L. Boyer. Optimization of importance factors in inverse planning. *Phys Med Biol*, 44(10):2525–2536, 1999.

[39] E.D. Yorke, G.J. Kutcher, A. Jackson, and C.C. Ling. Probability of radiation-induced complications in normal tissues with parallel architecture under conditions of uniform whole or partial organ irradiation. *Radiother Oncol*, 26(3):226–237, 1993.

Department of Radiation Oncology, University Hospitals of Cleveland, Case Western Reserve University School of Medicine, Cleveland, OH 44106

*E-mail address*: `qjw@po.cwru.edu`

Department of Mathematical Sciences, University of North Carolina, Greensboro, NC 27402

*E-mail address*: `wang@uncg.edu`

Department of Radiation Oncology, University Hospitals of Cleveland, Case Western Reserve University School of Medicine, Cleveland, OH 44106

DIMACS Series in Discrete Mathematics
and Theoretical Computer Science
Volume **55**, 2000

# Medical Applications of Discrete Tomography

Attila Kuba, Gabor T. Herman, Samuel Matej,
and Andrew Todd-Pokropek

ABSTRACT. Discrete tomography deals with a special type of tomographic inverse problem in which the function to be reconstructed from its projections is discrete (i.e., has only a few elements in its range of values). In the special case when the function is binary (only two possible values), prior information can be used to recover it from very few (e.g., only 2 or 3) projections.

While discrete tomography has a mathematical theory based mostly on discrete mathematics (in particular on combinatorial and discrete geometry) and has interesting theoretical problems concerning existence, uniqueness and algorithms in various discrete spaces, it also has some fascinating applications, many of which come from the field of medicine.

In this paper we review our own and related approaches to the medical applications of discrete tomography, such as reconstruction of human organs (for example, the heart or blood vessels) from 2 or 3 X-ray views as in biplane/triplane angiography, reconstruction from sparse radiographic data, and enhancement of classical tomographic images.

## 1. Introduction

We assume that there is a domain, which may itself be discrete (such as a set of ordered pairs of integers) or continuous (such as a Euclidean space). We further assume that there is an unknown function whose range is known to be a given discrete set (usually of real numbers). The problems of *discrete tomography*, as we perceive the field, have to do with determining the function (perhaps only partially, perhaps only approximately) from weighted sums over subsets of its domain in the discrete case and from weighted integrals over subspaces of its domain in the continuous case. In many applications these sums or integrals may be known only approximately. From this point of view, the most essential aspect of discrete tomography is that knowing the discrete range of the function may allow us to determine its value at points where without this knowledge it could not be determined. Discrete tomography is full of mathematically fascinating questions and it has many interesting applications.

---

2000 *Mathematics Subject Classification.* Primary 52C99, 05B20.

The authors wish to acknowledge the support of NSF/MTA joint grant No. 123, NSF grant DMS9612077, Hungarian Ministry of Education grant FKFP 0908, and OTKA grant T 0322421.

The reconstruction algorithms used in CT (computerized tomography; see, e.g., [**15**]) are derived from the not discrete (let us say, general) tomography model in which the range of the function to be reconstructed is the real numbers. Such reconstruction algorithms are unlikely to be appropriate for discrete tomography. For example, when using such an algorithm we cannot expect to produce a two-valued function, not even in the case when the data are taken from a two-valued function.

Since discrete functions can be considered to be special cases of general functions, discrete tomography can be thought of as a special kind of CT: it seems natural to apply the results of CT to discrete functions. However, it turns out that discrete tomography needs its own theory to answer questions concerning consistency, existence and uniqueness. Another reason for investigating special discrete reconstruction methods is that, since the function to be reconstructed is discrete, there is hope that it can be determined from less data than what are necessary for general functions. Accordingly, in discrete tomography the typical number of projections (a projection is a collection of line sums or line integrals for a set of lines which are either all parallel to each other or diverge from a single point) is two to four, which is much less than what is typically used in CT (a few hundred). General CT reconstruction methods cannot be used effectively if the number of projections is so small.

Discrete tomography has its own mathematical theory based mostly on discrete mathematics. It has strong connections with combinatorics and geometry. For a collection of results of discrete tomography see the Special Issue on Discrete Tomography of the *International Journal of Imaging Systems and Technology* [**16**] or the book *Discrete Tomography: Foundations, Algorithms, and Applications* [**17**].

In this paper we review some of the applications of discrete tomography to medicine. Our discussion is restricted to works which deal with discrete tomography as defined in the first paragraph of this section, and it is biased towards our own and related approaches. In the next section we discuss the modeling of medical objects in the context of discrete tomography. The projections of the object to be reconstructed are discussed in Section 3. The reconstruction methods are classified in Section 4 as heuristic methods, methods of solving linear equation system with binary constraints, and optimization methods. Finally, reported results on the medical applications of discrete tomography are surveyed in Section 5.

## 2. Modeling of objects

The methods of the discrete tomography can be used if the object to be reconstructed is *homogeneous* or consists of *a few homogeneous regions* having different physical characteristics. Even in the first case, there are two regions: the object and its background. For example, the object can be the patient's left ventricle filled with homogeneously distributed Röntgen contrast agent with the background consisting of the rest of the space, in which there is no dye. As an example of an object with several regions, we can take the human brain consisting of regions with different up-takes of a radio-pharmaceutical (representing the different functionality of the regions such as white and gray matter).

Typically, we use a function with a *discrete domain.* This domain consists of the points of some finite regular (2D or 3D) lattice. Let us denote the set of the possible values of the function by $D$, i.e., $D$ is the range of the function. We refer to such a function as a *D-image* (or just as an *image*, for short). The values of the image correspond to some physical property of the object; e.g., attenuation, density, etc. The image has the same value at all points of a single region and has different values at points in different regions. The number of values is equal to the number of regions in the object plus 1 (background). If we have a homogeneous object, then typically the object points are assigned the value 1 and the background points the value 0. Such a $\{0,1\}$-image is also referred to as a *binary image*.

In biomedical applications it is quite usual that a 3D image is considered to be a series of 2D slices and the slices are reconstructed one-by-one. The size of the lattice in one slice is typically between $64 \times 64$ and $512 \times 512$. In the slices the points are usually arranged in rows and columns.

We represent the image values at the lattice points by the vector $X = (x_j)$, with $x_j \in D$, for $1 \leq j \leq J$. It is typical in discrete tomography that the information available from the projections on its own is not sufficient for the unique determination of $X$. For this reason some *property*, known *a priori*, of the object is used during the reconstruction process.

- Such property can be that the object is similar to a given model. The model can be, for example, the averaged data set of representative objects; e.g., Onnasch and Prause used $64 \times 64 \times 64$ binary matrices [**27**] constructed from a collection of human studies for modeling the human ventricles in end-systole and end-diastole [**22**].
- Another frequently used property is that adjacent slices of the image are similar; more exactly: if there are many possible solutions consistent with the measurements taken of a slice, then we can try to find the solution which is the most similar to an adjacent (already reconstructed) slice (see, e.g., [**11, 27**]) or some smoothness criterion between the adjacent slices can be requested [**33**].
- There are cases when certain properties are explicitly required in the binary image to be reconstructed: for example, convexity in rows and columns (i.e., there is no 0 between two 1's) and/or 8-connectedness of the 2D slices [**29**], or in the case of objects such as vascular structures the aim can be to reconstruct in the slices elliptical objects [**28**], or, more generally, objects with smooth contours defined by spline functions [**33**].
- Finally, there can be some statistical information about the image. For example, the image can be a typical member of a class of images having a known Gibbs distribution

$$\Pi(X) = \frac{1}{Z} e^{-\beta H(X)} , \tag{2.1}$$

where $\Pi(X)$ is the probability of the occurrence of the image $X$, $Z$ is the normalizing factor, $\beta$ is a parameter of the distribution, and $H(X)$ is the so-called energy function. The energy function describes the characteristics of a typical image. It can express such properties as smoothness and shape of the borders of the object [**6, 7**] or it can encourage certain local configurations, e.g., uniform black or white clusters of pixels or corners [**3, 23**]. In this case the image can be considered to be a random sample from the known probability distribution.

In medical applications the reconstructed objects can be vascular structures [**31, 32, 33**], left and right heart ventricles [**26, 27**], brain slices [**6**], hemoglobin molecules [**10**], or proteins [**2**].

## 3. Projections

In order to get information about the object, measurements are performed, in most cases, along some rays. These rays can be transmitted through the object (e.g., in the case of X-rays or transmission electron microscopy) or emitted from the object (for example by some radioactive material distributed in it).

In an ideal situation, the measurements can be described mathematically as line integrals of the image along the rays. The measurements taken along a set of parallel lines constitute a *projection* of the object. The number of projections in discrete tomography is usually 2, 3, or 4. If two projections are collected, then the directions are almost always orthogonal to each other. Of course, in real-life measurements it is more realistic to take the integrals over strips or fan-beams instead of lines. In any case, the relation between the projection measurements and the image (containing the object) can be expressed as

$$y_i = \sum_{j=1}^{J} a_{ij} x_j, \qquad i = 1, 2, \ldots, I. \tag{3.1}$$

We use $Y = (y_i)$, where $y_i$ denotes the $i$-th measurement, and $A = (a_{ij})$, where $a_{ij}$ describes the contribution of the $j$-th discrete point to the $i$-th measurement. The values of $a_{ij}$ can be computed from a geometric model. For example, if the measurements are on strips, then $a_{ij}$ can be defined to be 1 if the $j$-th point is in the $i$-th strip, and to be 0 otherwise. The simplest case is when a binary image is to be reconstructed from two projections, namely from its row and column sums.

If the effects of noise and inexact measurements are taken into account then, instead of the linear equation system (3.1), we can consider the inequality system

$$|y_i - \sum_j a_{ij} x_j| \le \epsilon_i, \qquad i = 1, 2, \ldots, I, \tag{3.2}$$

where $\epsilon_i$ denotes an assumed upper bound of the error in the $i$-th measurement (see, e.g., [**31, 35**]).

Another way of handling noise is to use probabilistic models (see, e.g., [**6**]). For example, in the case of positron emission tomography (PET) the measurement $y_i$ can be considered to be a sample from a Poisson distribution whose mean is

$$l_i = \sum_j a_{ij} x_j, \qquad i = 1, 2, \ldots, I. \tag{3.3}$$

Then the probability of obtaining the measurement vector $Y$ if the image vector is $X$ is given by the likelihood function [**34**]

$$L(Y|X) = \prod_i \frac{l_i^{y_i} \cdot e^{-l_i}}{y_i!}. \tag{3.4}$$

## 4. Reconstruction methods

The basic problem in discrete tomography is that the information available from the projection measurements is generally not sufficient for the unique determination of the image. Usually the linear equation system (3.1) contains fewer equations than unknowns; i.e., $I < J$. The situation is only slightly better if we take into account the constraints $x_j \in D$, where $D$ is given. There are several methods to reconstruct binary images from their two projections; i.e., row and column sums (for an overview, see [**21**]).

In many cases a so-called *probability array* $W = (w_1, w_2, \ldots, w_J)$ is first computed in order to give some help with the reconstruction [**10, 25, 27, 29**]. Two published possibilities for $W$ are the *multiplicative probability array* ($w_j$ is the product of the projection measurements belonging to the projection lines going through the point $j$) and the *additive probability array* ($w_j$ is the sum of the projection measurements belonging to the projection lines going through the point $j$; this corresponds to the outcome of the back-projection operation of computerized tomography [**30**]).

Now we discuss the best-known discrete tomography reconstruction methods, classifying them into three classes.

**4.1. Heuristic methods.** The problem of multiple solutions can be handled by looking only for solutions satisfying some special properties. One of the earliest examples of this kind is when convex, axially symmetric polygons are reconstructed from two projections [**8**].

One might also attempt to find a solution from a small set of projections by using a probability array [**10**]. For example, using the multiplicative probability array, we can proceed as follows. We set to 1 one of the most probable points. Then we compute the new projections and the new probability array for the remaining points. This procedure can be repeated until the necessary number of 1's is found or the probability array contains only zeros. However this algorithm can fail even in very simple situations (see Fig. 1).

Another strategy is suggested by Onnasch *et al.* [**25, 26, 27**] to reconstruct 4D binary objects (moving heart ventricles, i.e., the fourth dimension is time). The 2D slices are reconstructed one by one using the information about the already reconstructed adjacent slices in time and space. More exactly, for the reconstruction of a 2D slice, we first compute the multiplicative probability array $W$ associated with that slice (as it is defined above). Instead of $W$, a *modified probability array*

$$\widetilde{W} = W \cdot T \cdot S$$

is used during the reconstruction, where $T$ and $S$ are blurred arrays of the temporally and spatially, respectively, adjacent and already reconstructed slices (in the case of the first slices the corresponding slices of a binary model can be used). Consequently, in the modified probability array $\tilde{w}_j$ has a high value if at the point $j$ the multiplicative probability array has a high value and the point $j$ is far in the object interior in the spatially and temporally adjacent slices. Then in the (initially all zero-valued) current slice we make two passes through all the points and we change a 0 to a 1 at point $j$ if and only if

$$y_i \geq \sum_j a_{ij} x_j \;, \qquad i = 1, 2, \ldots, I \;,$$

| (a) | | | | (b) | | | | (c) | | | |
|---|---|---|---|---|---|---|---|---|---|---|---|
| 0 | 1 | 0 | 0 | 0 | 2 | 1 | 0 | 0 | 1 | 1 | 0 |
| 1 | 0 | 0 | 1 | **16** | 0 | 0 | 2 | x | 0 | 0 | 1 |
| 1 | 0 | 0 | 0 | 2 | 0 | 0 | 1 | 1 | 0 | 0 | 1 |
| 0 | 0 | 1 | 0 | 0 | 1 | 2 | 0 | 0 | 1 | 1 | 0 |

FIGURE 1. Situation where the reconstruction method based on setting to 1 one of the most probable points failed. (a) The binary image to be reconstructed from the four projections taken from the horizontal, vertical and two diagonal directions. It is easy to check that this matrix is uniquely determined by the given four projections. (b) Initial (multiplicative) probability array. The most probable point is indicated by bold number. (c) Probability array after the first iteration. The already set point is indicated by x. There are points where the probability array has the maximal value, but if they are set to 1, then the process will fail to terminate in an image which satisfies the four projections.

remains valid. Roughly, during the first pass the most probable elements (i.e., with the greatest value of $\widetilde{w}_j$) are considered in an order from the center to the periphery of the 2D section and during the second pass only the points adjacent to ones which are already 1-valued are considered (to reduce the roughness of the slice contour).

**4.2. Solving linear equation systems with binary restriction.** It is a natural idea to try to find a binary solution of the system of linear equations (3.1), if it exists at all, or at least to find a binary vector near to a real-number-valued solution. The system (3.1) is usually solved using some (nonbinary) iterative method [**5**], for example, an Algebraic Reconstruction Technique (ART, [**12**]). In order to accommodate the binary restriction, Herman proposed a binary modification to ART (*BART* [**14**]). The idea behind BART was later extended by Censor and Matej [**4**] as the approach of binary steering of an arbitrary nonbinary iterative algorithm. To demonstrate this approach, we will utilize Cimmino's algorithm [**5**]. It starts from an initial image estimate $X^{(0)}$ (quite typically 0 is assigned to all points in the initial image). The new iterate $X^{(k+1)}$ is calculated from the previous iterate $X^{(k)}$ using the formula

$$X^{(k+1)} = X^{(k)} + \lambda_k \sum_{i=1}^{I} \omega_i^{(k)} \frac{y_i - \sum_j a_{ij} x_j^{(k)}}{\sum_j a_{ij}^2} a^i \, , \tag{4.1}$$

where $a^i$ denotes the $i$-th column of the transpose of the matrix $A$, $\lambda_k$ is a relaxation parameter used in the current iteration and the $\omega_i^k$ are positive iteration-dependent weights which must sum (over $i$) to one, for every $k$. If the initial $X^{(0)}$ was selected suitably, then the sequence $X^{(k)}$ converges to the minimal norm solution of the equation system (3.1), which is in general not binary.

The calculations within one iteration of the binary steering algorithm can be divided into the following three steps.

*Partial binarization.* Results of the previous iterate $X^{(k)}$ are thresholded to get an image $\widetilde{X}^{(k)}$ that has a 0 or 1 value at those points where the value in $X^{(k)}$ is "near" to 0 or 1. More precisely,

$$\widetilde{x}_j^{(k)} = \begin{cases} 0, & \text{if } x_j^{(k)} \le \alpha_k, \\ 1, & \text{if } x_j^{(k)} \ge \beta_k, \\ x_j^{(k)}, & \text{otherwise} , \end{cases}$$

for all $j \in J$, where the real sequences $\alpha_k$ and $\beta_k$, $0 \le \alpha_k < \beta_k \le 1$, denotes the thresholds applied in this iteration. (During the iterations $\alpha_k$ increases and $\beta_k$ decreases in order to get hopefully more and more $\{0,1\}$-valued points.)
*Correction application.* $\widetilde{X}^{(k)}$ is used (instead of $X^{(k)}$) for calculating the correction term of the nonbinary iterative algorithm. For example, if using Cimmino's algorithm, this step will be calculated using the formula

$$\widehat{X}^{(k+1)} = X^{(k)} + \lambda_k \sum_{i=1}^{I} \omega_i^k \frac{y_i - \sum_j a_{ij}\widetilde{x}_j^{(k)}}{\sum_j a_{ij}^2} a^i \ .$$

*Conflict settlement.* The new image estimate $X^{(k+1)}$ is generated from $\widehat{X}^{(k+1)}$ using the rule

$$x_j^{(k+1)} = \begin{cases} t_k - \epsilon, & \text{if } x_j^{(k)} \le \alpha_k \text{ and } \widehat{x}_j^{(k)} \ge t_k, \\ t_k + \epsilon, & \text{if } x_j^{(k)} \ge \beta_k \text{ and } \widehat{x}_j^{(k)} \le t_k, \\ \widehat{x}_j^{(k)}, & \text{otherwise}, \end{cases}$$

for all $j \in J$, where $\epsilon$ is a fixed real number (for example, $\epsilon = 0.01$ or $\epsilon = 0.05$), and $t_k$ is a threshold sequence (for example the sequence with the fixed value 0.5).

When a predetermined iteration index $K$ has been reached or the iterations are stopped at $K$ due to some stopping criteria, then a further thresholding is used to get the final binary image $X^*$ using

$$x_j^* = \begin{cases} 0, & \text{if } x_j^{(K)} \le 0.5, \\ 1, & \text{if } x_j^{(K)} > 0.5. \end{cases}$$

Simulation results for BART and the binary steering approach are presented in [**14**] and [**4**], respectively.

**4.3. Optimization.** The reconstruction problem can also be posed as one of optimization. We give four examples.

4.3.1. *Finding similar slices.* The idea in this case is to minimize a cost function expressing similarity with a model (which can be, in the case of a slice-by-slice 3D or 4D reconstruction, the previous slice) under the constraint of (3.1). The cost function can be expressed as

$$\sum_j c_j x_j \tag{4.2}$$

where the components of the cost vector $C$ can be defined in the most simple case as

$$c_j = \begin{cases} 0, & \text{if the point } j \text{ has value 1 in the model,} \\ v, & \text{otherwise,} \end{cases}$$

where $v$ denotes some fixed positive value. A similar idea, but a more complex cost vector, was introduced in [**31**] and [**35**] in such a way that the value of $c_j$ was higher at points further from the contour of the object in the model. To find a minimizer of (4.2) under the constraints of (3.1), we can rewrite it as a minimal cost flow problem [**35**], which can be solved, for example, by Klein's algorithm [**20**]. If (3.2) is used for modeling the measurements and (4.2) is to be minimized [**31, 35**], a modified version of the Busacker and Gowen algorithm can be used [**18**].

4.3.2. *Finding a binary image best matching the given projections.* If the given projections are noisy, we can try to find a binary image having projections nearest to the given projection data. In this case the cost function to be minimized may look like

$$C = \|Y - AX\|^2 .$$

One way of obtaining an optimal $X$ is provided by the method of "simulated annealing" [**19**]. It starts the iterations with an initial estimate of the binary image, denoted by $X^{(0)}$. Since the number of 1's of the binary images $X^{(k)}$ will remain the same during the iterations, the initial estimate $X^{(0)}$ should be created such that it contains as many points with value 1 as the binary image to be reconstructed. The number of 1's should be estimated from the projections. In each iteration step, one point of $X^{(k)}$ will be changed from 1 to 0 and another one will be changed from 0 to 1, providing a new image $X'$. An attempted change is accepted or rejected according to the criterion proposed in [**24**]. Consider the difference of the values of the cost function $C$ between the two images $X'$ and $X^{(k)}$:

$$\delta C = \|Y - AX'\|^2 - \|Y - AX^{(k)}\|^2 .$$

The image $X'$ will be accepted, i.e., $X^{(k+1)} = X'$, if and only if $e^{-\delta C/\lambda T}$ is greater than a random number generated from a uniform distribution between 0 and 1, where $\lambda$ is a suitably selected positive constant and $T$ is a positive control parameter (resembling "temperature" in statistical mechanics). It is clear that a change is accepted for any $T$ if $\delta C \leq 0$. However, if $\delta C > 0$, then the change will be accepted with a certain probability which is greater if $T$ is greater. The value of $T$ is high at the beginning of the algorithm and it is slowly decreased during the iteration steps. The stopping criterion of the iterations can be that the number of accepted changes falls under some preset fraction of the number of attempted changes. This method was used in [**32**] to reconstruct 3D binary images from multiple-view cone-beam projections.

4.3.3. *Finding a smooth 3D binary image best matching the given projections - combined optimization.* It is also possible to pose an optimization problem by combining the previous two ideas (interslice resemblance and data fidelity) with others (e.g., contour regularity) and to minimize a more complex cost function such as [**28, 33**]

$$C = \|Y - AX\|^2 + \lambda_1 \Phi_1(X) + \lambda_2 \Phi_2(X) .$$

Here $\lambda_1$ and $\lambda_2$ are positive regularization coefficients, $\Phi_1(X)$ measures the chamfer distances [**1**] between contours in adjacent slices of $X$ and $\Phi_2(X)$ measures the irregularity of the slice contours of $X$. Simulated annealing can also be used to get a solution for such functions [**33**].

4.3.4. *Finding a binary solution with Bayes estimators.* If we are looking for a most probable image $X$ knowing the projections $Y$ and supposing some $\Pi(X)$ distribution of the possible images, then the function to be optimized can be, for example, the *a posteriori probability*

$$P(X|Y) = \frac{\Pi(X) \cdot L(Y|X)}{P(Y)} ,$$

where $L(Y|X)$ is the probability of obtaining $Y$ if the image is $X$. This has the same maximum as

$$\Pi(X) \cdot L(Y|X) . \tag{4.3}$$

In the case of PET, and assuming a Gibbs prior, maximizing (4.3) is reduced to the following problem on the basis of (2.1), (3.3) and (3.4): find the $X$ which minimizes

$$\sum_i \Big( \sum_j a_{ij}x_j - y_i \cdot \log(\sum_j a_{ij}x_j) \Big) + \beta H(X) .$$

The optimal solution can again be found by simulated annealing [**6**]. If, instead of (3.3) and (3.4) we assume a Gaussian model for the measurement error, then the function to be minimized in order to find the *maximum a posteriori probability* (MAP) estimator becomes

$$\alpha\|Y - AX\|^2 + \beta H(X) .$$

Another Bayes estimator can be the *minimum mean square error* (MMSE) estimate [**6**], defined by

$$E[X|Y] = \sum P(X|Y) \cdot X .$$

## 5. Results

In order to measure the difference between the original and the reconstructed binary images, $X$ and $X'$, respectively, several measures of the errors have been suggested, such as the *relative mean error* [**9, 13**].

$$\overline{R} = \frac{\sum |x_j - x'_j|}{\sum x_j} \cdot 100 \ \% ,$$

the *percentage of the misplaced voxels* [**32**]

$$MV = \frac{\sum |x_j - x'_j|}{2 \cdot \sum x_j} \cdot 100 \ \%$$

(note that, $MV = \overline{R}/2$), the *mean square error* [**6, 7**], the *chamfer distance* [**1**], the *shape*, and *volume errors*

$$SE = \frac{2 \cdot \sum |x_j - x'_j|}{\sum x_j + \sum x'_j} \cdot 100 \ \% \quad \text{and} \quad VE = \frac{2 \, |\sum x_j - \sum x'_j|}{\sum x_j + \sum x'_j} \cdot 100 \ \% ,$$

and the so-called *shape* and *volume conformities* [**27, 29, 35**])

$$SC = 100\,\% - SE \qquad \text{and} \qquad VC = 100\,\% - VE.$$

The performance of the different reconstruction methods was tested on software and physical phantoms and in clinical studies.

**5.1. Software phantom studies.** One of the first software phantom studies was performed by Herman [**13**] to test his BART method by reconstructing various artificial $25 \times 25$ binary images from 2 and 4 projections. The relative mean error was between 0% and 57% depending on the shape of the object to be reconstructed and on the noise added to the projections.

Onnasch and Heintzen did reconstructions from two computer-generated projections of synthetic ideal ventriculograms representing diastolic and systolic human hearts [**25**]. They used a method of reconstructing slice by slice by making use of similarity to a model in time and space. They found a conformity of more than 94% between the actual and the reconstructed objects.

Software phantom studies were performed to reconstruct a $128 \times 128$ slice of a hemoglobin molecule from 3 projections [**10**] by the heuristic method described in Section 4.1. No error estimations were given in this case.

Slump and Gerbrands [**35**] tried their reconstruction method for a software phantom generated from 10 slices of the left ventricle of a dog at 8 phases in the cardiac cycle. The method was based on an optimization procedure with the previously reconstructed adjacent phase as the model. The relative mean error varied between 19.4% and 71.6%.

Robert and co-workers [**32**] simulated X-ray angiographic cone-beam projections of binary images. They used the optimization reconstruction method which finds a solution which best matches the given projections. They demonstrated that this method allows one to reconstruct a sphere and a simulated branched blood vessel from three views with $MV = 3\%$ and 7%, respectively.

Pellot and coworkers [**28**] tested their reconstruction method based on combined optimization (see Subsection 4.3.3). They used two projections taken from two arbitrary directions. The objects to be reconstructed were various 2D and 3D simulated blood vessels. They demonstrated that their method performs acceptably both on single and on branching vessels.

Two MAP image reconstruction methods were tested by Chan and co-workers [**6**] on a PET phantom image of size $95 \times 95$. They applied a two-stage reconstruction procedure:
1. Perform a reconstruction using a filtered backprojection algorithm to produce an initial image.
2. Perform a Bayesian restoration of the initial image to produce the final image.

They compared various priors using the mean square error to measure differences between images.

Cardiac angiographic phantoms were created by Carvalho and co-workers [**3**]. The phantoms consisted of mathematically-described approximations of the left and right ventricles and the left atrium (see Fig. 2), the image sizes were $63 \times 63$. Three projections were generated (from the horizontal, the vertical, and a diagonal direction). As reconstruction method a modification of simulated annealing was used. The average $MV$ error was between 1.3% and 3.1% depending on the simulated noise level (see Fig. 2).

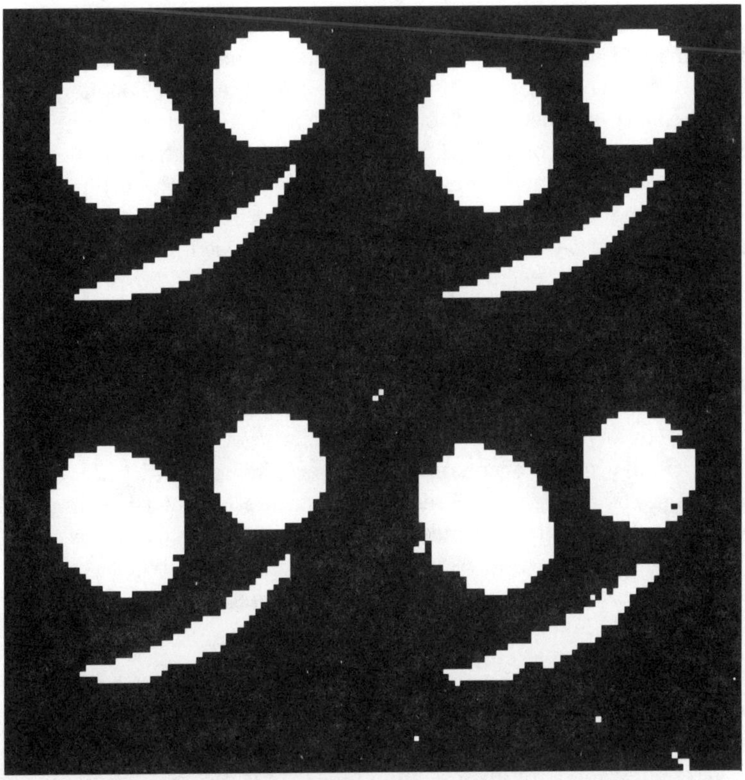

FIGURE 2. A phantom (upper left corner) and its reconstructions using three noiseless projections (upper right corner), using low noise projections (lower left corner) and high noise projections (lower right corner); further figures reporting on similar experiments appear in [**3**].

Very recently Sensali and coworkers [**33**] published a paper on the reconstruction of blood vessels from X-ray angiograms. The method is based on combined optimization (see Subsection 4.3.3). They studied the influence of several parameters (e.g., regularization parameters, SNR) on reconstruction errors and demonstrated that quite complex shapes can be acceptably reconstructed.

**5.2. Physical phantom studies.** Chang and Chow tried their heuristic reconstruction method on a clay model of a dog heart [**8**]. X-ray films were taken from two orthogonal directions and digitized. After some preprocessing steps (logarithmic transformation, boundary detection, smoothing, registration, scaling) the reconstruction was performed and the reconstructed slices displayed. No quantitative results of error estimations were reported.

As a preliminary evaluation of their reconstruction procedure, Reiber and coworkers performed a physical phantom test [**31**]. They made a perspex test model, having circular cross sections with diameters ranging from 4 mm to 1.5 mm, representing a coronary arterial segment with some obstruction. The model was filled with contrast medium and filmed in two orthogonal projections. They tested three reconstruction methods creating 290 $64 \times 64$ slices and obtained the best results with Onnasch's method [**25**], for which the relative mean error was 13.2%.

Van Tran and coworkers [**36**] tried their moment inversion reconstruction method on a lucite stenosis phantom and on a segment of a coronary artery of an *in vitro* heart taking biplane X-ray images. They compared the areas in the reconstructed cross-sections with the areas in the slices of the artery.

For validation of their binary reconstruction method, Prause and Onnasch performed several physical phantom studies [**29**]. They used a small dye-filled plastic bottle as the phantom object. They collected two $256 \times 256$ projections and reconstructed a $128 \times 128 \times 128$ 3D image from them. They could reach shape and volume conformities of 84.4% and 96.4%, respectively. A more realistic test object (a left- and a right-ventricular phantom) was modeled from plasticine. The convex-shaped left-ventricular phantom was reconstructed with 87.3% shape and 98.0% volume conformities with a reconstruction method which did not use of a model. The right-ventricular phantom was reconstructed with 75.8% shape and 96.0% volume conformities with a reconstruction method using a simple model.

Eccentric and concentric cylindrical stenosis phantoms were used to test the method of Senasli *et al.* [**33**]. They were able to reconstruct the global shape correctly. The diameters of the slices of the phantoms could be estimated from the images with acceptable accuracy.

**5.3. Clinical studies.** One of the first clinical studies was reported by Reiber and his co-workers [**31**]. A part of the right coronary artery was reconstructed from two cineangiograms by applying Slump's method [**35**].

The system developed by Prause and Onnasch for reconstruction of heart chambers from biplane angiographic projections [**29**] is currently undergoing clinical validation. It is able to create $128 \times 128 \times 128$ 3D images of the heart chambers at different phases (including end-diastole and end-systole). Sophisticated preprocessing steps are necessary to get the projection data from the angiograms (logarithmic conversion, subtraction of structures not belonging to the chambers, geometric and densitometric corrections, and image pairing). They applied the reconstruction method of [**25**] which uses similar slices in time and space and binary models of averaged casts of human ventricles [**22**]. Absolute volume data were derived from the reconstructed 3D binary images and compared with other methods of volume determination.

## 6. Summary

In this paper we have surveyed much of the currently available literature on the medical applications of discrete tomography, as defined in the first paragraph of the Introduction. We have discussed how objects are modeled by $D$-images and the relationship of such images to their projection data. We have given an overview of algorithms which appeared in literature for the reconstruction of biomedical objects from their projections under the assumption that the range of values of the object to be reconstructed is a (small) discrete set. Finally, we have reported on experiments which have been performed to evaluate the efficacy of such algorithms.

## References

[1] G. Borgefors, *Distance transformations in arbitrary dimensions*, Comput. Vision Graphics Image Process. vol. 27, pp. 321–345, 1984.

[2] J.M. Carazo, C.O. Sorzano, E. Reitzel, R. Schröder, and R. Marabini, *Discrete tomography in electron microscopy*, in G.T. Herman, A. Kuba (Eds.), Discrete Tomography. Foundations, Algorithms, and Applications, Birkhäuser: Boston, MA, 1999, pp. 405–416.

[3] B.M. Carvalho, G.T. Herman, S. Matej, C. Salzberg, and E. Vardi, *Binary tomography for triplane cardiography*, in A. Kuba, M. Samal, and A. Todd-Poropek (Eds.), Information Processing in Medical Imaging, LNCS-1613, Springer-Verlag: Berlin, Germany, 1999, pp. 29–41.

[4] Y. Censor, S. Matej, *Binary steering of nonbinary iterative algorithms*, in G.T. Herman, A. Kuba (Eds.), Discrete Tomography. Foundations, Algorithms, and Applications, Birkhäuser: Boston, MA, 1999, pp. 285–296.

[5] Y. Censor and S.A. Zenios, *Parallel Optimization: Theory, Algorithms, and Applications*, Oxford University Press: New York, NY, 1997.

[6] M.T. Chan, G.T. Herman, and E. Levitan, *Bayesian image reconstruction using image-modeling Gibbs priors*, Int. J. Imaging Systems Techn. vol. 9, pp. 85–98, 1998.

[7] M.T. Chan, G.T. Herman, E. Levitan, *Probabilistic modeling of discrete images*, in G.T. Herman, A. Kuba (Eds.), Discrete Tomography. Foundations, Algorithms, and Applications, Birkhäuser: Boston, MA, 1999, pp. 213–235.

[8] S.-K. Chang and C.K. Chow, *The reconstruction of three-dimensional objects from two orthogonal projections and its application to cardiac cineangiography*, IEEE Trans. Comp. vol. C-22, pp. 18–28, 1973.

[9] S.-K. Chang and G.L. Shelton, *Two algorithms for multiple-view binary pattern reconstruction*, IEEE Trans. Systems, Man, Cybernetics vol. SMC-1, pp. 90–94, 1971.

[10] A.V. Crewe and D.A. Crewe, *Inexact reconstruction: Some improvements*, Ultramicroscopy vol. 16, pp. 33–40, 1985.

[11] J.J. Gerbrands and C.H. Slump, *3-D reconstruction of homogeneous objects from two Poisson-distributed projections*, Pattern Recognition Letters vol. 3, pp. 137–145, 1985.

[12] R. Gordon, R. Bender, and G.T. Herman, *Algebraic Reconstruction Techniques (ART) for three-dimensional electron microscopy and X-ray photography*, J. Theor. Biol. vol. 29, pp. 471–481, 1970.

[13] G.T. Herman, *Two direct methods for reconstructing pictures from their projections: A comparative study*, Comp. Graphics Image Proc. vol. 1, pp. 123–144, 1972.

[14] G.T. Herman, *Reconstruction of binary patterns from a few projections*, in A. Günther, B. Levrat, H. Lipps (Eds.), Intern. Computing Symp., North-Holland: Amsterdam, The Netherlands, 1974, pp. 371–378.

[15] G.T. Herman, *Image Reconstruction from Projections: The Fundamentals of Computerized Tomography*, Academic Press: New York, NY, 1980.

[16] G.T. Herman and A. Kuba (Eds.), *Special Issue: Discrete Tomography*, Int. J. Imaging Systems Techn. vol. 9., No. 2/3, 1998.

[17] G.T. Herman and A. Kuba (Eds.), *Discrete Tomography: Foundations, Algorithms and Applications*, Birkhäuser: Boston, MA, 1999.

[18] T. C. Hu, *Integer Programming and Network Flows*, Addison-Wesley: Reading, Mass., 1969.

[19] S. Kirkpatrick, C. D. Gelatt, M. P. Vechi, *Optimization by simulated annealing*, Science vol. 220, pp. 671–680, 1983.

[20] M. Klein, *A primal method for minimal cost flows with applications to the assignment and transportation problem*, Manag. Sci. vol. 14, pp. 205–220, 1967.

[21] A. Kuba and G. T. Herman, *Discrete tomography: A historical overview*, in G.T. Herman and A. Kuba (Eds.), Discrete Tomography. Foundations, Algorithms, and Applications, Birkhäuser: Boston, MA, 1999, pp. 3–34.

[22] P. E. Lange, D. G. W. Onnasch, F. L. Farr, V. Malerczyk, and P. Heintzen, *Analysis of left- and right ventricular size ans shape, as determined from human casts: Description of the method and its validation*, Europ. J. Cardiol. vol 8, pp. 431–448, 1978.

[23] S. Matej, A. Vardi, G. T. Herman and E. Vardi, *Binary tomography using Gibbs priors*, in G.T. Herman, A. Kuba (Eds.), Discrete Tomography. Foundations, Algorithms, and Applications, Birkhäuser: Boston, MA, 1999, pp. 191–212.

[24] N. Metropolis, A. Rosenbluth, M. Rosenbluth, A. Teller, and E. Teller, *Equation of state calculation by fast computing machines*, J. Chem. Phys. vol. 21, pp. 1087–1092, 1953.

[25] D.G.W.Onnasch and P.H. Heintzen, *A new approach for the reconstruction of the right or left ventricular form from biplane angiographic recordings*, in Computers in Cardiology, IEEE Comp. Soc.: Long Beach, 1976, pp. 67–73.

[26] D.G.W. Onnasch, W. Schmitz, P.H. Heintzen, *Problems of the binary reconstruction of the left and right ventricles from biplane angiocardiograms*, in P. H. Heintzen, R. Brennecke (Eds.), Digital Imaging in Cardiovascular Radiology, Georg Thieme Verlag: Stuttgart, Germany, 1983, pp. 141–151.

[27] D.G.W. Onnasch and G.P.M. Prause, *Heart chamber reconstruction from biplane angiography*, in G.T. Herman, A. Kuba (Eds.), Discrete Tomography. Foundations, Algorithms, and Applications, Birkhäuser: Boston, MA, 1999, pp. 385–403.

[28] C. Pellot, A. Herment, M. Sigelle, P. Horain, H. Maitre, and P. Peronneau, *A 3D reconstruction of vascular structures from two X-ray angiograms using an adapted simulated annealing algorithm*, IEEE Trans. Medical Imaging vol. 15, pp. 532–559, 1996.

[29] G.P.M. Prause and D.G.W. Onnasch, *Binary reconstruction of the heart chambers from biplane angiographic image sequence*, IEEE Trans. Medical Imaging vol. 13, pp. 48–60, 1994.

[30] G.N. Ramachandran and A.V. Lakshminarayanan, *Three-dimensional reconstruction from radiographs and electron micrographs: Application of convolutions instead of Fourier transforms*, Proc. Natl. Acad. Sci. USA vol. 68, pp. 2236–2240, 1971.

[31] J.H.C. Reiber, J.J. Gerbrands, G.J. Troost, C.J. Kooijman, C.H. Slump, *3-D reconstruction of coronary arterial segments from two projections*, in P. H. Heintzen, R. Brennecke (Eds.), Digital Imaging in Cardiovascular Radiology, Georg Thieme Verlag: Stuttgart, Germany, 1983, pp. 151–163.

[32] N. Robert, F. Peyrin, and M.J. Yaffe, *Binary vascular reconstruction from a limited number of cone beam projections*, Med. Phys. vol. 21, pp. 1839–1851, 1994.

[33] M. Senasli, L. Garnero, A. Herment, and E. Mousseaux, *3D reconstruction of vessel lumen from very few angiograms by dynamic contours using a stochastic approach*, Graphical Models vol. 62, pp. 105–127, 2000.

[34] L.A. Shepp and Y. Vardi, *Maximum likelihood reconstruction in emission tomography*, IEEE Trans. Medical Imaging vol. MI-1, pp. 113–121, 1982.

[35] C.H. Slump and J.J. Gerbrands, *A network flow approach to reconstruction of the left ventricle from two projections*, Comp. Graphics Image Proc. vol. 18, pp. 18–36, 1982.

[36] L. Van Tran, R.C. Bahn, and J. Sklansky, *Reconstructing the cross sections of coronary arteries from biplane angiograms*, IEEE Trans. Medical Imaging vol. MI-12, pp. 517–529, 1992.

Applied Informatics Department, University of Szeged, Szeged, Hungary
*E-mail address*: kuba@inf.u-szeged.hu

Radiology Department, University of Pennsylvania, Philadelphia, PA, USA
*E-mail address*: gabor@mipg.upenn.edu

Radiology Department, University of Pennsylvania, Philadelphia, PA, USA
*E-mail address*: matej@mipg.upenn.edu

Medical Physics and Bio-Engineering Department, University College, London, UK
*E-mail address*: atoddpok@medphys.ucl.ac.uk

DIMACS Series in Discrete Mathematics
and Theoretical Computer Science
Volume **55**, 2000

# A Study of 3D Imaging Approaches in Medicine

Jayaram K. Udupa

## 1. Introduction

In medical imaging, there are many sources of multidimensional images [**CJS93**]; 2D: digital/digitized radiographic images, tomographic slices; 3D: a time sequence of 2D images of a dynamic object, a stack of slice images of a static object; 4D: a time sequence of 2D images of a dynamic object for a range of imaging parametric values (e.g., MR spectroscopic images of a heart), a time sequence of 3D images of a dynamic object; 5D: a time sequence of 3D images of a dynamic object for a range of parametric values. Three- and higher-dimensional images may also be generated by a computational step. For example, a 5D binary image is produced if we apply a "lifting" operation [**UH99**] to a 4D gray-level image to represent the 4D image intensity distribution as a surface in a 5D space. For ease of further reference, we will refer to a multidimensional image simply as a *scene* and represent it by a pair $\mathcal{C} = (C, f)$, where $C$, the *scene domain*, is a multidimensional rectangular array of spatial elements (*spels* for short), and $f$, the *scene intensity*, is a function that assigns to every spel a vector whose component elements are from a set of integers.

There is usually an *object* of study for which the scene is generated. This may be a physical object such as an organ/tissue component, a tumor/fracture/lesion/an abnormality, a prosthetic device, a phantom, or a conceptual object such as an isodose surface in a radiation treatment plan, an activity region highlighted by PET or fMRI, a mathematical phantom. The objects may be rigid, deformable, static, or dynamic. We will refer to any operation, which, for a given set of multimodality scenes of an object of study, produces information about the object of study, as *3D imaging*. The information may be qualitative, for example a 3D rendition of the object, or quantitative, for example, the volume of the object. The purpose of 3D imaging is, therefore, given multiple scenes as input, to output information about the object of study.

In this exposition, we will give a quick overview of the various 3D imaging operations that are commonly used and identify the challenges currently faced. We argue with strong evidence that, since the scenes are inherently digital, we should take entirely digital approaches to realize all 3D imaging operations. We also suggest that, since object information in scenes is always fuzzy, we should take fuzzy approaches. The need to fulfill these strategies open numerous discrete

The author's research is supported by an NIH grant NS37172, a grant from the Department of Army DAMD 179717271, and a contract from EPIX Medical, Inc.

problems in both hard and fuzzy settings. The reader is referred to [**UH99**] for a detailed account of 3D imaging operations and their medical applications.

## 2. An overview of 3D imaging operations

3D imaging operations may be classified into four groups: *preprocessing*, *visualization*, *manipulation*, and *analysis*. All preprocessing operations aim at improving or extracting object information from the given scenes. The purpose of visualization operations is to assist humans in perceiving and comprehending the object structure/characteristics/function in two, three and higher dimensions. The manipulation operations allow humans to interactively alter the object structures (mimic surgery). The analysis operations enable us to quantify object structural/morphological/functional information. In the rest of this section, we will cursorily examine these four classes of operations. The references given here are only samples from a literature which has perhaps over a thousand publications excluding applications.

**2.1. Preprocessing operations.** The input to these operations is a set of scenes of a body region and the output is either a set of scenes or a *structure system* – a computer representation of the objects imaged in the scene. Usually the structure system contains only one structure corresponding to the object of interest.

Volume of Interest (VOI) [**Udu82**]: The input scene $\mathcal{C}_i = (C_i, f_i)$ is converted to another smaller scene $\mathcal{C}_o = (C_o, f_o)$ such that $C_o \subset C_i$ and $f_o$ is a restriction of $f_i$ to $C_o$. $C_o$ is specified, usually interactively, to be the set of spels in a rectangular box, a sphere, or an ellipsoid (and their equivalents in higher dimensions). The purpose here is to make $\mathcal{C}_o$ just contain the objects of interest and to reduce the amount of data to be handled in other expensive operations. In large, routine applications, to automate this operation is a challenge. For this, clearly, some object knowledge is essential.

Filtering: In these operations, the output scene $\mathcal{C}_o$ is obtained from the input scene $\mathcal{C}_i$ by modifying the intensity values in such a way as to either *enhance* object information [**PM83**] or to *suppress* unwanted non-object information such as noise [**CY83**] in $\mathcal{C}_i$. Usually for these operations $C_o = C_i$. Various types of filters are available under these two categories. Unfortunately, usually some non-object information is also enhanced by the former methods and object information is also suppressed by the latter methods. The challenge in filtering is how to incorporate object knowledge into the filter design so as to achieve optimum performance.

Interpolation: The purpose of interpolation is to change the level and orientation of discretization of $\mathcal{C}_i$ to a desired entity. For these operations, usually $C_o \neq C_i$. These operations are needed for converting a non isotropically sampled scene into a scene with isotropic resolution, or for obtaining scenes at desired level and orientation of discretization. Two classes of operations are available: scene-based and object-based. In *scene-based* methods [**HRY79**], $f_o(c)$ is estimated from $f_i(c')$ for spels $c'$ lying in the close vicinity of $c$. In *object-based* methods [**RU90**], some object information is used in guiding interpolation. Object-based methods have been shown repeatedly in the literature to produce better results than scene-based methods. The challenge here is how to incorporate object knowledge specific to the application into the interpolation process.

Registration: The purpose of registration is to represent structures (object representations) obtained from multiple scenes in a common coordinate system. These are needed for combining object information obtained from multiple modalities (such as MRI, PET and CT) for the same body region and for determining change, growth, motion, displacement of objects over time. Again two classes of methods are available: scene-based and object-based. In *scene-based* methods [**WCM92**], scene intensity patterns are matched, while in *object-based* methods [**PCS$^+$89**] structures derived from scenes are matched. Both rigid and deformable methods of registration under both categories are available. As in other operations, the challenge here is to incorporate specific object knowledge into the registration process in both object-based and scene-based methods.

Segmentation: This operation, the most crucial perhaps among all 3D imaging operations, outputs a structure system from a given set of scenes. It consists of two related tasks – recognition and delineation. *Recognition* is the process of determining roughly the whereabouts of the objects in the scene, and *delineation* is the process of determining precisely the spatial extent and composition of the objects in the scene. Knowledgeable humans usually outperform computer algorithms in the high-level task of recognition, while computer algorithms can outperform humans in the precise, accurate, and efficient delineation of objects.

There are two classes of approaches to recognition – automatic/knowledge-based/model-based [**GK95**] and human-assisted. By far the latter is the most commonly used in practice. At the outset, two classes of approaches to delineation may be identified – *boundary-based* [**MT95**] and *region-based* [**US96**]. In the former, the output is a set of boundaries of the object, and in the latter, it is a set of object regions. Each of these two strategies can be further divided into subgroups – hard [**MT95**] and fuzzy [**US96**] – depending on whether the output is a hard or a fuzzy set. Combined with the two strategies of recognition, therefore, we may identify eight classes of approaches to segmentation.

As seen from the outline of other preprocessing operations, object knowledge usually facilitates all 3D imaging operations, including segmentation. This implies that segmentation is helpful/needed for all 3D imaging operations, ironically also for segmentation itself. To devise generic segmentation methods, with high precision, accuracy and efficiency, that can be quickly adapted to a given application, is indeed the greatest challenge in 3D imaging and in image analysis.

**2.2. Visualization.** The input to these operations is a set of scenes or a structure system and the output is a *rendition* – a 2D scene that is a visual portrayal of the object information contained in the input scenes/structures. Two classes of methods are available: *scene-based* wherein renditions are created directly from given scenes, and *object-based* wherein renditions are created from (hard or fuzzy) structures extracted from the scene. Scene-based methods may be further divided into two groups: slice mode and volume mode. In the former, 2D "slices" of different orientations – natural, orthogonal, oblique and curved – are extracted from the scene and displayed as a montage or in a roam through fashion using gray scale, color or overlay [**UH99**]. In the latter, surfaces, interfaces and intensity distributions are displayed with a variety of 3D cues using *surface rendering* and *volume rendering* techniques [**GR87, Lac95**]. In object-based methods, the hard or fuzzy structures extracted from the scene are displayed with 3D cues using surface rendering and volume rendering techniques [**UO91, UO93**]. The challenges in visualization are the realistic display of medical objects including color, texture and surface properties, speeding up volume rendering, and objective evaluation of the large number of rendering schemes available.

**2.3. Manipulation.** These operations are needed for simulating surgery, for unobscured visualization, and for developing aids for interventional procedures. They take as input a structure system and output another structure system by altering structures or their relationship. Two classes of operations are being developed: rigid [**UO91, OU95**] with operations to cut, separate, add, subtract, move and mirror structures and their components and deformable [**Wat87**]. The main challenge in manipulation is to realize the manipulative operations realistically utilizing tissue material properties.

**2.4. Analysis.** The purpose of these operations is, given a set of scenes or a structure system, to generate a quantitative description of the morphology or function of the objects in the body region captured in the scene. Two classes of operations are available: *scene-based* wherein quantities based on scene intensities such as intensity statistics within object regions, tissue density, activity, perfusion, flow are estimated, and *object-based* wherein quantities measured from segmented structures and how they change with time are estimated including distance, length, width, curvature, area, volume, kinematics and mechanics. The main challenge here is the assessment of the accuracy of the estimated measures.

## 3. Go digital, go fuzzy

Any scene of any body region exhibits two important characteristics of the objects contained in the body region.

*Graded Composition*: The spels in the same object region exhibit a heterogeneity of scene intensity due to the heterogeneity of the object material and noise, blurring, and background variation introduced by the imaging device. Even if a perfectly homogeneous object were to be imaged, its scene will exhibit graded composition.

*Hanging-togetherness (Gestalt)*: In spite of the graded composition, knowledgeable humans do not have any difficulty in perceiving object regions as a whole (Gestalt). This is a fuzzy phenomenon and should be captured through a proper theoretical and computational framework. That is, there are no binary objects or acquired binary scenes. Measured data always have uncertainties. Additionally, scenes are inherently digital.

As seen from the previous section, no matter what 3D imaging operation we consider, we cannot ignore these two fundamental facts. We have to deal with essentially two types of data – scenes and structures. We argue that, instead of imposing some sort of a continuous model on the scene or the structure in a hard fashion, taking an inherently digital approach, preferably in a fuzzy setting, for all 3D imaging operations can lead to effective, efficient and practically viable methods. Taking such an approach would require, in almost all cases, the development of the necessary mathematical theories and algorithms from scratch. We need appropriate theories and algorithms for topology, geometry and morphology, all in a fuzzy, digital setting, for dealing with the concept of a "structure" in scenes. Note that almost all the challenges we raised in the previous section relate to some form of object definition in scenes. Therefore, they all need the above mathematical and algorithmic developments. Additionally, since we deal with deformable objects (all soft-tissue organs, and often even bone, as in distraction osteogenesis), we also need fuzzy digital mechanics theories and algorithms to handle structure data realistically for the operations relating to registration, segmentation, manipulation and analysis. Because of the digital setting, almost all challenges raised previously lead to discrete problems. Since we are dealing with $n$-dimensional scenes, the mathematics and algorithms need to be developed for hard and fuzzy sets of spels defined in $n$ dimensions.

Before concluding this section, we will give two examples of digital/fuzzy approaches that motivated us to the above argument. The first relates to modeling an object as a 3D surface and visualizing it. The second relates to the fuzzy topological concept of connectedness and its use in segmentation.

**3.1. Polygonal versus digital surfaces.** In hard, boundary-based 3D scene segmentation (e.g., by thresholding), the output structure is often a 3D surface. The surface is represented usually either as a set of polygons (most commonly triangles [**LC87**]), or in a digital form [**USH82**] as a set of cubes or as a set of oriented faces of cubes. ("Oriented" here means that a face with a normal vector pointing from inside of the surface to its outside in the $+x$ direction is distinguished from a face at the same location with a face normal from inside to outside pointing in the $-x$ direction.) Usually the surface normals at various points on these surfaces (in both the polygonal and the digital case) are estimated independent of the geometry of the surface elements, for example, by the gradient of the intensity function $f$ of the original scene from which the surface was segmented [**HB86**]. Since the surface normals dictate the shading in the rendering process, the surface elements are used here only as a geometric guide rather than as detailed shape descriptors of the surface. Since the surface elements in the digital case are simple *and* are all of identical size and shape, they can be stored using clever data structures that are typically an order of magnitude more compact than their polygonal counterparts [**GU99**]. Further, there is a well developed body of literature (e.g., [**KR96**], Chapters 6, 7) describing the theory of digital surfaces that naturally generalizes to $n$-dimensions for any $n$.

More importantly, rendering of digital surfaces is considerably simpler and more efficient than that of polygonal surfaces, the main reason being the simplicity of the surface elements and of their spatial arrangement in the former case [**UO91**]. It was shown in [**GU99**] that, digital surface rendering entirely in software on a 300 MHz Pentium PC can be done about 4–17 times faster than the rendering of the same objects, whose surfaces are represented by triangles, on a Silicon Graphics Reality II hardware rendering engine for about the same quality of renditions. The enormous speed advantage for digital surfaces comes from the fact that, since the surface elements are all of the same shape and size, most of the expensive computations can be reduced to table lookup operations. It is also possible to exploit the special geometry of the spatial arrangement of the digital surface elements to simplify hidden surface computations considerably.

**3.2. Fuzzy connected object definition.** Although much work has been done in digital geometry and topology based on hard sets, analogous work based on fuzzy sets is rare [**Ros79, US96**]. As pointed out earlier, the uncertainties about objects inherent in scenes must be retained as realistically as possible in all operations instead of making arbitrary hard decisions. Although many fuzzy strategies have been proposed particularly for image segmentation, almost all of them have not considered the spatio-topological relationship among spels in a fuzzy setting to formulate the concept of hanging-togetherness. This is a vital piece of information that can greatly improve the immunity of object definition (segmentation) methods to noise, blurring and background variation. Given the fuzzy nature of object information in scenes, the frameworks to handle fuzziness in scenes should address questions of the following form: How are objects to be mathematically defined in a fuzzy, digital setting taking into account the graded composition and hanging-togetherness of spels? How are fuzzy boundaries to be defined satisfying a Jordan boundary property? What are the appropriate algorithms to extract these entities from scenes in such a way as to satisfy the definitions? These questions are largely open. We will give one example below of the type of approaches that can be taken. This relates to fuzzy connected object definition [**US96**]. This framework and the algorithms have now been applied extensively on 1000's of scenes in several clinical applications [**UWS+97, UOT+97, SUCC99**].

We define a fuzzy *adjacency* relation $\alpha$ on spels independent of the scene intensities. The strength of this relation is in $[0, 1]$ and is greater when the spels are spatially closer. The purpose of $\alpha$ is to capture the blurring property of the imaging device. We define another fuzzy relation $\kappa$, called *affinity*, on spels. The strength of this relation between any two spels $c$ and $d$ lies in $[0, 1]$ and depends on $\alpha$ as well as on how similar are the scene intensities and other properties derived from scene intensities in the vicinity of $c$ and $d$. Affinity is a local fuzzy relation. If $c$ and $d$ are far apart, their affinity is 0. Fuzzy *connectedness* is yet another fuzzy relation on spels, defined as follows. For any two spels $c$ and $d$ in the scene, consider all possible connecting paths between them. (A path is simply a sequence of spels such that the successive spels in the sequence are "nearby".) Every such path has a strength which is the smallest affinity of successive pairwise spels along the path. The strength of fuzzy connectedness between $c$ and $d$ is the largest of the strength of all paths between $c$ and $d$. A fuzzy connected object of a certain strength $\theta$ is a pool $O$ of spels together with an objectness value assigned to each spel in $O$. $O$ is such that for any spels $c$ and $d$ in $O$, their strength of connectedness is at least $\theta$, and for any spels $e$ in $O$ and $g$ not in $O$, their strength is less than $\theta$. Although the computation of a fuzzy connected object in a given scene for a given $\kappa$ and $\theta$ appears to be combinatorially explosive, the theory leads to solutions for this problem based on dynamic programming. In fact, fuzzy connected objects in 3D scenes ($256 \times 256 \times 60$) can be extracted at interactive speeds (a few seconds) on modern PCs such as a 400 MHz Pentium PC.

## 4. Concluding remarks

In this article, we have first given an overview of the operations available for 3D imaging in medicine – a discipline wherein given a set of multidimensional scenes the aim is to extract, visualize, manipulate, and analyze object information captured in the scenes. We have also raised numerous challenges that are encountered in real applications. We have presented an argument with evidence that there are considerable advantages in taking an inherently digital approach to realizing all 3D imaging operations. We have also argued that, since object information in scenes is fuzzy, the digital approaches should be developed in a fuzzy framework to handle the uncertainties realistically. This calls for the development of topology, geometry, morphology and mechanics, all in a fuzzy and digital setting, all of which are likely to have a significant impact on medical imaging and the practice of medicine.


## Acknowledgments

The author is grateful to Mary A. Blue for typing the manuscript.


## References


[CJS93] Z. H. Cho, J. P. Jones, and M. Singh, *Foundations of medical imaging*, John Wiley & Sons, Inc., New York, NY, 1993.

[CY83] R. T. Chin and C. L. Yeh, *Quantitative evaluation of some edge-preserving noise smoothing techniques*, Comput. Graph. Image Process. **23** (1983), 67–91.

[GK95] L. Gong and C. Kulikowski, *Composition of image analysis processes through object-oriented hierarchical planning*, IEEE Trans. Patt. Anal. Machine Intell. **17** (1995), no. 10, 997–1008.

[GR87] S. Goldwasser and R. Reynolds, *Real-time display and manipulation of 3-D medical objects: The voxel machine architecture*, Comput. Vision Graph. Image Process. **39** (1987), no. 4, 1–27.

[GU99] G. J. Grevera and J. K. Udupa, *Order of magnitude faster surface rendering via software in a PC than using dedicated hardware*, SPIE Proc., vol. 3658, 1999, pp. 202–211.

[HB86] K. H. Höhne and R. Bernstein, *Shading 3D images from CT using gray-level gradients*, IEEE Trans. Med. Imaging **5** (1986), 45–47.

[HRY79] G. T. Herman, S. W. Rowland, and M.-M. Yau, *A comparative study of the use of linear and modified cubic spline interpolation for image reconstruction*, IEEE Trans. Nucl. Med. **26** (1979), no. 2, 2879–2894.

[KR96] T. Y. Kong and A. Rosenfeld (eds.), *Topological algorithms for digital image processing*, Elsevier Science B. V., New York, NY, 1996.

[Lac95] P. G. Lacroute, *Fast volume rendering using a shear-warp factorization of the viewing transformation*, Tech. Report CSL-TR-95-678, Ph.D. Thesis, Departments of Electrical Engineering and Computer Science, Stanford University, Stanford, California, September 1995.

[LC87] W. Lorensen and H. Cline, *Marching cubes: A high resolution 3D surface construction algorithm*, Comput. Graph. **21** (1987), 163–169.

[MT95] T. McInerney and D. Terzopoulos, *A dynamic finite element surface model for segmentation and tracking in multidimensional medical images with application to cardiac 4D image analysis*, Comput. Med. Imaging Graph. **19** (1995), no. 1, 69–83.

[OU95] D. Odhner and J. K. Udupa, *Shell manupulation: Inteeeractive alteration of multiple-material fuzzy structures*, SPIE Proc., vol. 2431, 1995, pp. 35–42.

[PCS+89] C. A. Pelizzari, G. T. Y. Chen, D. R. Spelbring, R. R. Weichselbaum, and C.-T. Chen, *Accurate three-dimensional registration of CT, PET, and/or MR images of the brain*, J. Comput. Assist. Tomogr. **13** (1989), 20–26.

[PM83] P. Perona and J. Malik, *Scale space and edge detection using anisotropic diffusion*, IEEE Trans. Patt. Anal. Machine Intell. **12** (1983), 629–639.

[Ros79] A. Rosenfeld, *Fuzzy digital topology*, Inform. Control **40** (1979), 76–87.

[RU90] S. P. Raya and J. K. Udupa, *Shape-based interpolation of multidimensional objects*, IEEE Trans. Med. Imaging **9** (1990), no. 1, 32–42.

[SUCC99] P. K. Saha, J. K. Udupa, E. F. Conant, and D. P. Chakraborty, *Near-automatic segmentation and quantification of mammographic glandular tissue density*, SPIE Proc., vol. 3661, 1999, pp. 266–276.

[Udu82] J. K. Udupa, *Interactive segmentation and boundary surface formation for 3-D digital images*, Comput. Graph. Image Process. **18** (1982), 213–235.

[UH99] J. K. Udupa and G. T. Herman (eds.), *3D imaging in medicine*, 2nd ed., CRC Press, Boca Raton, FL, 1999.

[UO91] J. K. Udupa and D. Odhner, *Fast visualization, manipulation, and analysis of binary volumetric objects*, IEEE Comput. Graph. Applicat. **11** (1991), no. 6, 53–62.

[UO93] J. K. Udupa and D. Odhner, *Shell rendering*, IEEE Comput. Graph. Applicat. **13** (1993), no. 6, 58–67.

[UOT+97] J. K. Udupa, D. Odhner, J. Tian, G. Holland, and L. Axel, *Automatic clutter-free volume rendering for MR angiography using fuzzy connectedness*, SPIE Proc., vol. 3034, 1997, pp. 111–119.

[US96] J. K. Udupa and S. Samarasekera, *Fuzzy connectedness and object definition: Theory, algorithms, and applications in image segmentation*, Graph. Models Image Process. **58** (1996), no. 3, 246–261.

[USH82] J. K. Udupa, S. N. Srihari, and G. T. Herman, *Boundary detection in multidimensions*, IEEE Trans. Patt. Anal. Machine Intell. **4** (1982), 41–50.

[UWS+97] J. K. Udupa, L. Wei, S. Samarasekera, Y. Miki, M. A. van Buchem, and R. I. Grossman, *Multiple sclerosis lesion quantification using fuzzy connectedness principles*, IEEE Trans. Med. Imaging **16** (1997), 246–261.

[Wat87] K. Waters, *A muscle model for animating three dimensional facial expression*, Proc. SIGGRAPH'87 **21** (1987), no. 3, 17–24.

[WCM92] R. P. Woods, S. R. Cherry, and J. C. Mazziotta, *Rapid automated algorithm for aligning and reslicing PET images*, J. Comput. Assist. Tomogr. **17** (1992), no. 4, 620–633.

Medical Image Processing Group, Department of Radiology, University of Pennsylvania Health Systems, Philadelphia, Pennsylvania 19104

*E-mail address*: `jay@mipg.upenn.edu`

DIMACS Series in Discrete Mathematics
and Theoretical Computer Science
Volume **55**, 2000

# Discrete Mathematics in Medical Imaging: a personal view

Peter Hall

## 1. Introduction

The DIMACS workshop associated with this volume carried a panel session. The panel discussed principally, but not wholly, the value of purely digital approaches to medical image processing, one panelist (Professor Udupa) advocated that we should "go digital!". This I found to be a very interesting proposition, and this short paper gives my thoughts on the matter.

## 2. Discussion

I begin with the observation that much of the research literature pertinent to medical image processing is premised on the following assertion: that the digital images we commonly use as input are pictures of continuous real world bodies, such as livers, hearts, brains, bones, blood vessels, and so on. That this is so can be demonstrated quite easily by glancing at any of the journals or conference proceedings. This should not be a surprise, it seems intuitive and obvious to model those real world bodies using continuous mathematics, and regard the digital image as an approximation. The panel discussion challenged this orthodox view, and proposed that we explicitly use discrete mathematics at all times, on the grounds that we have digital (discrete) images as input, the "digital proposition", as it might be called. This proposition suggests that efficient digital approaches are possible, where previously continuous approach were thought necessary.

It is true that working wholly in the discrete domain will bring a whole new set of research questions, so the proposition is interesting. However, rather than explore what these questions might be, I would like to step back a little and ask about the feasibility of the proposition. I will use examples to show not only that continuous and discrete representations should be carefully chosen for a task, if efficiency is to be maximised, but also that there are tasks for which only one approach will do. Hence neither a purely digital nor purely continuous approach is possible.

Let us consider a typical task in medical imaging: that of comparing two like organs from different people, the liver say. This comparison involves constructing a mapping between spatial points in each liver. It seems to me that continuous approaches are sure to dominate over digital approaches in tasks such as this, despite the fact that (currently) we begin with discrete representations of the livers. The reason is that we are faced with a continuous problem — the organs are continuous and the mapping is also continuous. Furthermore, any discrete representation is sure to be less compact than a corresponding continuous representation, because the discrete approach comprises a set of points whereas the continuous approach is able to exploit relationships between points. This may affect algorithmic efficiency to such a degree that efficient digital algorithms are not possible in practice (but they may be more efficient in theory).

The above example suggests that continuous representations will almost certainly lead to more efficient algorithms for matching than digital representations. However, it may be the case that there is an efficient digitally-based algorithm, but we just do not know it yet. Thus we must ask "is there a task we might want to perform that *necessitates* the use of continuous representations?" The answer is "yes".

To argue this I turn to an example from my own work, in which I am given several instances of a blood vessel, and must represent those instances in such a way as to recognise novel instances in new images. I describe vessel geometry with an n-dimensional vector, which I interpret as a point in n-dimensional space. Notionally, each distinct vessel is a new point in this space, and because vessel geometry varies continuously these points are continuously distributed. We have several instances of geometry, and so can model the continuous distribution (I use a Gaussian). Continuity is needed to compute new points from the distribution, that is to say new geometric instances. This is generalisation, and is useful in image processing because it allows us to match novel vessel geometries when when appear in new images.

Now we consider where digital approaches are beneficial or necessary, again via examples. It is true that set theoretic operations can be very difficult using B-rep models, which here we regard as a continuous representation. However, such operations are highly efficient on binary digital images. Furthermore, we argue that digital images are necessary if an arbitrary picture is to be stored on a computer. This is because how to explicitly represent any object in any picture is currently a question no one can answer. However, digital images provide an implicit representation (that is, they are data but not information) and hence represent any object, subject to size and frequency constraints.

This short article has followed the panel's topic quite faithfully in that it has concentrated on images, until now. Nowhere have I mentioned the idea of proving algorithms to be correct, or the most efficient, or some other property. However, there were two presentations of such a nature at the workshop (by Professor Wang). For such proofs the classical forms of discrete mathematics must be used. Such presentations are a rarity in the medical imaging literature — any proofs they do contain tend to be for mathematical results, rather than algorithm — the medical imaging algorithms tend to tested rather than proven. Whether this is necessary or merely cultural I cannot say.

## 3. Conclusion

Should we accept the "digital proposition?". My answer is "no", and adopt instead a pragmatic view: use which representation is best for the given task, bearing in mind that some tasks require an exclusive representation. For example, in my own work I use both discrete and continuous forms: I model a set of vasculature, using discrete approaches to model branching structure and its variations, and continuous approaches to model geometry and its variations. Both approaches allow for generalisation, but in different kinds of ways: with discrete forms I generalise, essentially, by permuting elements between different sets, but with the continuous approach I model absent members of a single set.

Does it matter that researchers in medical images tend to assume continuous representations? I think that the panel debate showed it does matter, if only to make us think more before we chose a particular representation: the "digital proposition" has, for me, brought certain assumptions into the open, and made me question them explicitly. That I arrived at a rather benign, pragmatic conclusion should not disguise the usefulness of the exercise (it was pointed out to me by a psychologist that it can be useful to have "common sense" confirmed).

What should be done? There is clearly room for research to push digital representations into domains currently dominated by continuous forms. If done well this could lead to efficiency gains, but in my view there will always be problems for which continuous forms are more appropriate. Perhaps one fruitful avenue for formalism might be to help us decide what the boundaries of utility are for any particular representation.

MATHEMATICAL SCIENCES, BATH UNIVERSITY, ENGLAND
*E-mail address*: `maspmh@maths.bath.ac.uk`

# Selected Titles in This Series

*(Continued from the front of this publication)*

25 **Gilbert Baumslag, David Epstein, Robert Gilman, Hamish Short, and Charles Sims, Editors,** Geometric and Computational Perspectives on Infinite Groups

24 **Louis J. Billera, Curtis Greene, Rodica Simion, and Richard P. Stanley, Editors,** Formal Power Series and Algebraic Combinatorics/Séries Formelles et Combinatoire Algébrique, 1994

23 **Panos M. Pardalos, David I. Shalloway, and Guoliang Xue, Editors,** Global Minimization of Nonconvex Energy Functions: Molecular Conformation and Protein Folding

22 **Panos M. Pardalos, Mauricio G. C. Resende, and K. G. Ramakrishnan, Editors,** Parallel Processing of Discrete Optimization Problems

21 **D. Frank Hsu, Arnold L. Rosenberg, and Dominique Sotteau, Editors,** Interconnection Networks and Mapping and Scheduling Parallel Computations

20 **William Cook, László Lovász, and Paul Seymour, Editors,** Combinatorial Optimization

19 **Ingemar J. Cox, Pierre Hansen, and Bela Julesz, Editors,** Partitioning Data Sets

18 **Guy E. Blelloch, K. Mani Chandy, and Suresh Jagannathan, Editors,** Specification of Parallel Algorithms

17 **Eric Sven Ristad, Editor,** Language Computations

16 **Panos M. Pardalos and Henry Wolkowicz, Editors,** Quadratic Assignment and Related Problems

15 **Nathaniel Dean and Gregory E. Shannon, Editors,** Computational Support for Discrete Mathematics

14 **Robert Calderbank, G. David Forney, Jr., and Nader Moayeri, Editors,** Coding and Quantization: DIMACS/IEEE Workshop

13 **Jin-Yi Cai, Editor,** Advances in Computational Complexity Theory

12 **David S. Johnson and Catherine C. McGeoch, Editors,** Network Flows and Matching: First DIMACS Implementation Challenge

11 **Larry Finkelstein and William M. Kantor, Editors,** Groups and Computation

10 **Joel Friedman, Editor,** Expanding Graphs

9 **William T. Trotter, Editor,** Planar Graphs

8 **Simon Gindikin, Editor,** Mathematical Methods of Analysis of Biopolymer Sequences

7 **Lyle A. McGeoch and Daniel D. Sleator, Editors,** On-Line Algorithms

6 **Jacob E. Goodman, Richard Pollack, and William Steiger, Editors,** Discrete and Computational Geometry: Papers from the DIMACS Special Year

5 **Frank Hwang, Fred Roberts, and Clyde Monma, Editors,** Reliability of Computer and Communication Networks

4 **Peter Gritzmann and Bernd Sturmfels, Editors,** Applied Geometry and Discrete Mathematics, The Victor Klee Festschrift

3 **E. M. Clarke and R. P. Kurshan, Editors,** Computer-Aided Verification '90

2 **Joan Feigenbaum and Michael Merritt, Editors,** Distributed Computing and Cryptography

1 **William Cook and Paul D. Seymour, Editors,** Polyhedral Combinatorics